These safety symbols are used in laboratory and field investigations in this book to indicate possible hazards. Learn the meaning of each symbol and refer to this page often. *Remember to wash your hands thoroughly after completing lab procedures.*

PROTECTIVE EQUIPMENT   Do not begin any lab without the proper protection equipment.

GOGGLES	Proper eye protection must be worn when performing or observing science activities that involve items or conditions as listed below.	**APRON** Wear an approved apron when using substances that could stain, wet, or destroy cloth.	**SOAP** Wash hands with soap and water before removing goggles and after all lab activities.	**GLOVES** Wear gloves when working with biological materials, chemicals, animals, or materials that can stain or irritate hands.

LABORATORY HAZARDS

Symbols	Potential Hazards	Precaution	Response
DISPOSAL	contamination of classroom or environment due to improper disposal of materials such as chemicals and live specimens	• DO NOT dispose of hazardous materials in the sink or trash can. • Dispose of wastes as directed by your teacher.	• If hazardous materials are disposed of improperly, notify your teacher immediately.
EXTREME TEMPERATURE	skin burns due to extremely hot or cold materials such as hot glass, liquids, or metals; liquid nitrogen; dry ice	• Use proper protective equipment, such as hot mitts and/or tongs, when handling objects with extreme temperatures.	• If injury occurs, notify your teacher immediately.
SHARP OBJECTS	punctures or cuts from sharp objects such as razor blades, pins, scalpels, and broken glass	• Handle glassware carefully to avoid breakage. • Walk with sharp objects pointed downward, away from you and others.	• If broken glass or injury occurs, notify your teacher immediately.
ELECTRICAL	electric shock or skin burn due to improper grounding, short circuits, liquid spills, or exposed wires	• Check condition of wires and apparatus for fraying or uninsulated wires, and broken or cracked equipment. • Use only GFCI-protected outlets	• DO NOT attempt to fix electrical problems. Notify your teacher immediately.
CHEMICAL	skin irritation or burns, breathing difficulty, and/or poisoning due to touching, swallowing, or inhalation of chemicals such as acids, bases, bleach, metal compounds, iodine, poinsettias, pollen, ammonia, acetone, nail polish remover, heated chemicals, mothballs, and any other chemicals labeled or known to be dangerous	• Wear proper protective equipment such as goggles, apron, and gloves when using chemicals. • Ensure proper room ventilation or use a fume hood when using materials that produce fumes. • NEVER smell fumes directly. • NEVER taste or eat any material in the laboratory.	• If contact occurs, immediately flush affected area with water and notify your teacher. • If a spill occurs, leave the area immediately and notify your teacher.
FLAMMABLE	unexpected fire due to liquids or gases that ignite easily such as rubbing alcohol	• Avoid open flames, sparks, or heat when flammable liquids are present.	• If a fire occurs, leave the area immediately and notify your teacher.
OPEN FLAME	burns or fire due to open flame from matches, Bunsen burners, or burning materials	• Tie back loose hair and clothing. • Keep flame away from all materials. • Follow teacher instructions when lighting and extinguishing flames. • Use proper protection, such as hot mitts or tongs, when handling hot objects.	• If a fire occurs, leave the area immediately and notify your teacher.
ANIMAL SAFETY	injury to or from laboratory animals	• Wear proper protective equipment such as gloves, apron, and goggles when working with animals. • Wash hands after handling animals.	• If injury occurs, notify your teacher immediately.
BIOLOGICAL	infection or adverse reaction due to contact with organisms such as bacteria, fungi, and biological materials such as blood, animal or plant materials	• Wear proper protective equipment such as gloves, goggles, and apron when working with biological materials. • Avoid skin contact with an organism or any part of the organism. • Wash hands after handling organisms.	• If contact occurs, wash the affected area and notify your teacher immediately.
FUME	breathing difficulties from inhalation of fumes from substances such as ammonia, acetone, nail polish remover, heated chemicals, and mothballs	• Wear ... and gloves. • Ens... or use a fume hood w... fumes. • N...	• If a spill occurs, leave area and notify your teacher immediately.
IRRITANT	irritation of skin, mucous membranes, or respiratory tract due to materials such as acids, bases, bleach, pollen, mothballs, steel wool, and potassium permanganate	• •	• If skin contact occurs, immediately flush the affected area with water and notify your teacher.
RADIOACTIVE	excessive exposure from alpha, beta, and gamma particles	• Remove gloves and wash ... water before removing remainder o... equipment.	• If cracks or holes are found in the container, notify your teacher immediately.

D1529291

Your online portal to everything you need

 **ConnectED**

connectED.mcgraw-hill.com

Look for these icons to access
exciting digital resources

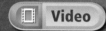 Video

 Audio

Review

Inquiry

WebQuest

Assessment

Concepts in Motion

FROM BACTERIA TO PLANTS

SCIENCE

Glencoe

FROM BACTERIA
TO PLANTS

iSCIENCE

Glencoe

Snow Leopard, *Uncia uncia*

The snow leopard lives in central Asia at altitudes of 3,000 m–5,500 m. Its thick fur and broad, furry feet are two of its adaptations that make it well suited to a snowy environment. Snow leopards cannot roar but can hiss, growl, and make other sounds.

The McGraw·Hill Companies

 Education

Send all inquiries to:
McGraw-Hill Education
8787 Orion Place
Columbus, OH 43240-4027

ISBN: 978-0-07-888014-8
MHID: 0-07-888014-9

Printed in the United States of America.

6 7 8 9 10 LWI 18 17

Authors

American Museum of Natural History
New York, NY

Michelle Anderson, MS
Lecturer
The Ohio State University
Columbus, OH

Juli Berwald, PhD
Science Writer
Austin, TX

John F. Bolzan, PhD
Science Writer
Columbus, OH

Rachel Clark, MS
Science Writer
Moscow, ID

Patricia Craig, MS
Science Writer
Bozeman, MT

Randall Frost, PhD
Science Writer
Pleasanton, CA

Lisa S. Gardiner, PhD
Science Writer
Denver, CO

Jennifer Gonya, PhD
The Ohio State University
Columbus, OH

Mary Ann Grobbel, MD
Science Writer
Grand Rapids, MI

Whitney Crispen Hagins, MA, MAT
Biology Teacher
Lexington High School
Lexington, MA

Carole Holmberg, BS
Planetarium Director
Calusa Nature Center and
Planetarium, Inc.
Fort Myers, FL

Tina C. Hopper
Science Writer
Rockwall, TX

Jonathan D. W. Kahl, PhD
Professor of Atmospheric Science
University of Wisconsin-
Milwaukee
Milwaukee, WI

Nanette Kalis
Science Writer
Athens, OH

S. Page Keeley, MEd
Maine Mathematics and
Science Alliance
Augusta, ME

Cindy Klevickis, PhD
Professor of Integrated Science
and Technology
James Madison University
Harrisonburg, VA

Kimberly Fekany Lee, PhD
Science Writer
La Grange, IL

Michael Manga, PhD
Professor
University of California, Berkeley
Berkeley, CA

Devi Ried Mathieu
Science Writer
Sebastopol, CA

Elizabeth A. Nagy-Shadman, PhD
Geology Professor
Pasadena City College
Pasadena, CA

William D. Rogers, DA
Professor of Biology
Ball State University
Muncie, IN

Donna L. Ross, PhD
Associate Professor
San Diego State University
San Diego, CA

Marion B. Sewer, PhD
Assistant Professor
School of Biology
Georgia Institute of Technology
Atlanta, GA

Julia Meyer Sheets, PhD
Lecturer
School of Earth Sciences
The Ohio State University
Columbus, OH

Michael J. Singer, PhD
Professor of Soil Science
Department of Land, Air and
Water Resources
University of California
Davis, CA

Karen S. Sottosanti, MA
Science Writer
Pickerington, Ohio

Paul K. Strode, PhD
I.B. Biology Teacher
Fairview High School
Boulder, CO

Jan M. Vermilye, PhD
Research Geologist
Seismo-Tectonic Reservoir
Monitoring (STRM)
Boulder, CO

Judith A. Yero, MA
Director
Teacher's Mind Resources
Hamilton, MT

Dinah Zike, MEd
Author, Consultant,
Inventor of Foldables
Dinah Zike Academy;
Dinah-Might Adventures, LP
San Antonio, TX

Margaret Zorn, MS
Science Writer
Yorktown, VA

Consulting Authors

Alton L. Biggs
Biggs Educational Consulting
Commerce, TX

Ralph M. Feather, Jr., PhD
Assistant Professor
Department of Educational
Studies and Secondary
Education
Bloomsburg University
Bloomsburg, PA

Douglas Fisher, PhD
Professor of Teacher Education
San Diego State University
San Diego, CA

Edward P. Ortleb
Science/Safety Consultant
St. Louis, MO

Series Consultants

Science

Solomon Bililign, PhD
Professor
Department of Physics
North Carolina Agricultural
and Technical State University
Greensboro, NC

John Choinski
Professor
Department of Biology
University of Central Arkansas
Conway, AR

Anastasia Chopelas, PhD
Research Professor
Department of Earth and
Space Sciences
UCLA
Los Angeles, CA

David T. Crowther, PhD
Professor of Science Education
University of Nevada, Reno
Reno, NV

A. John Gatz
Professor of Zoology
Ohio Wesleyan University
Delaware, OH

Sarah Gille, PhD
Professor
University of California
San Diego
La Jolla, CA

David G. Haase, PhD
Professor of Physics
North Carolina State
University
Raleigh, NC

Janet S. Herman, PhD
Professor
Department of Environmental
Sciences
University of Virginia
Charlottesville, VA

David T. Ho, PhD
Associate Professor
Department of Oceanography
University of Hawaii
Honolulu, HI

Ruth Howes, PhD
Professor of Physics
Marquette University
Milwaukee, WI

**Jose Miguel Hurtado, Jr.,
PhD**
Associate Professor
Department of Geological
Sciences
University of Texas at El Paso
El Paso, TX

Monika Kress, PhD
Assistant Professor
San Jose State University
San Jose, CA

Mark E. Lee, PhD
Associate Chair & Assistant
Professor
Department of Biology
Spelman College
Atlanta, GA

Linda Lundgren
Science writer
Lakewood, CO

Series Reviewers, continued

Carolyn Elliott
Iredell-Statesville Schools
Statesville, NC

Christine M. Jacobs
Ranger Middle School
Murphy, NC

Jason O. L. Johnson
Thurmont Middle School
Thurmont, MD

Felecia Joiner
Stony Point Ninth Grade Center
Round Rock, TX

Joseph L. Kowalski, MS
Lamar Academy
McAllen, TX

Brian McClain
Amos P. Godby High School
Tallahassee, FL

Von W. Mosser
Thurmont Middle School
Thurmont, MD

Ashlea Peterson
Heritage Intermediate Grade Center
Coweta, OK

Nicole Lenihan Rhoades
Walkersville Middle School
Walkersvillle, MD

Maria A. Rozenberg
Indian Ridge Middle School
Davie, FL

Barb Seymour
Westridge Middle School
Overland Park, KS

Ginger Shirley
Our Lady of Providence Junior-Senior High School
Clarksville, IN

Curtis Smith
Elmwood Middle School
Rogers, AR

Sheila Smith
Jackson Public School
Jackson, MS

Sabra Soileau
Moss Bluff Middle School
Lake Charles, LA

Tony Spoores
Switzerland County Middle School
Vevay, IN

Nancy A. Stearns
Switzerland County Middle School
Vevay, IN

Kari Vogel
Princeton Middle School
Princeton, MN

Alison Welch
Wm. D. Slider Middle School
El Paso, TX

Linda Workman
Parkway Northeast Middle School
Creve Coeur, MO

Teacher Advisory Board

The Teacher Advisory Board gave the authors, editorial staff, and design team feedback on the content and design of the Student Edition. They provided valuable input in the development of *Glencoe ①Science.*

Online Guide

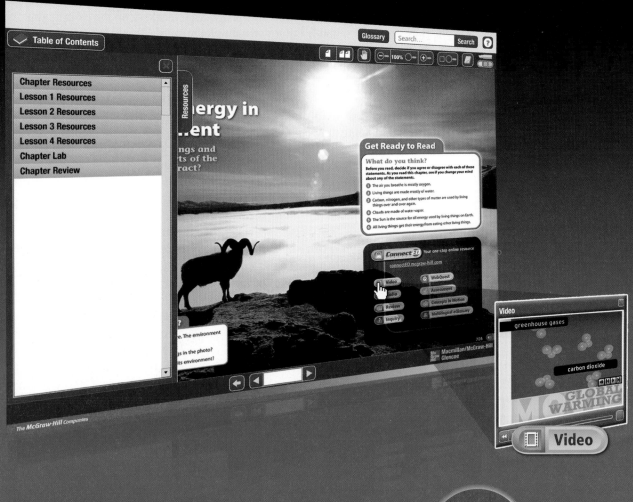

ConnectED

▷ **Your Digital Science Portal**

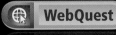

Video	Audio	Review	Inquiry	WebQuest
See the science in real life through these exciting videos.	Click the link and you can listen to the text while you follow along.	Try these interactive tools to help you review the lesson concepts.	Explore concepts through hands–on and virtual labs.	These web-based challenges relate the concepts you're learning about to the latest news and research.

The icons in your online student edition link you to interactive learning opportunities. Browse your online student book to find more.

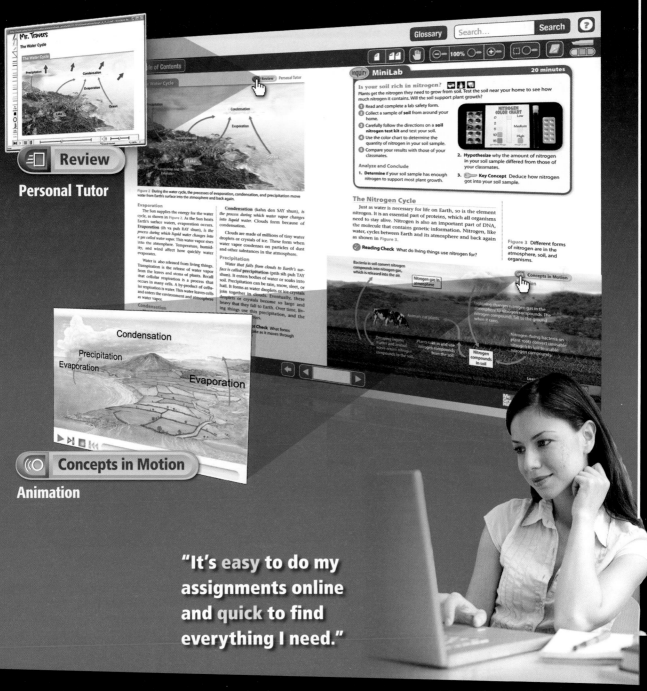

Review

Personal Tutor

Concepts in Motion

Animation

"It's easy to do my assignments online and quick to find everything I need."

✓ **Assessment**

Check how well you understand the concepts with online quizzes and practice questions.

(○ **Concepts in Motion**

The textbook comes alive with animated explanations of important concepts.

g **Multilingual eGlossary**

Read key vocabulary in 13 languages.

Treasure Hunt

Your science book has many features that will aid you in your learning. Some of these features are listed below. You can use the activity at the right to help you find these and other special features in the book.

-  **THE BIG IDEA** can be found at the start of each chapter.

- The Reading Guide at the start of each lesson lists 🔑 **Key Concepts**, vocabulary terms, and online supplements to the content.

- **ConnectED** icons direct you to online resources such as animations, personal tutors, math practices, and quizzes.

- **Inquiry** Labs and Skill Practices are in each chapter.

- Your **FOLDABLES®** help organize your notes.

1 What four margin items can help you build your vocabulary?

2 On what page does the glossary begin? What glossary is online?

3 In which Student Resource at the back of your book can you find a listing of Laboratory Safety Symbols?

4 Suppose you want to find a list of all the Launch Labs, MiniLabs, Skill Practices, and Labs, where do you look?

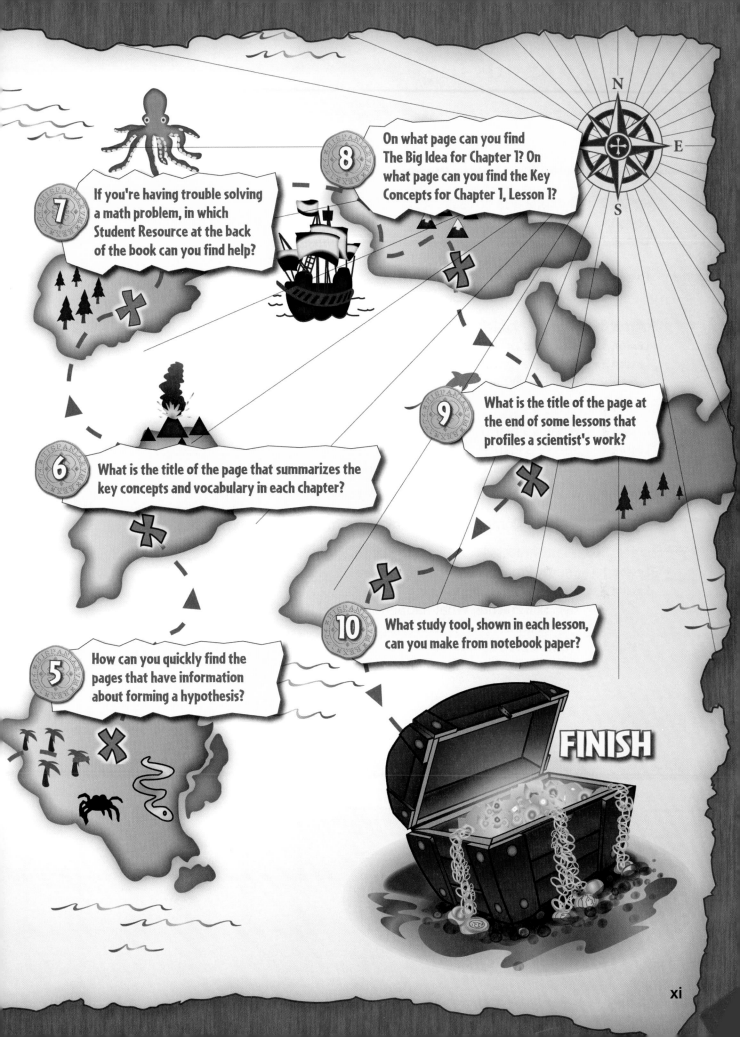

On what page can you find The Big Idea for Chapter 1? On what page can you find the Key Concepts for Chapter 1, Lesson 1?

8

If you're having trouble solving a math problem, in which Student Resource at the back of the book can you find help?

7

What is the title of the page at the end of some lessons that profiles a scientist's work?

9

What is the title of the page that summarizes the key concepts and vocabulary in each chapter?

6

What study tool, shown in each lesson, can you make from notebook paper?

10

How can you quickly find the pages that have information about forming a hypothesis?

5

FINISH

Table of Contents

Unit 2 **From Bacteria to Plants** 224

Chapter 7 **Bacteria and Viruses** 228
Lesson 1 What are bacteria? 230
Lesson 2 Bacteria in Nature...238
  (Inquiry) **Skill Practice** How do lab techniques affect an investigation? 245
Lesson 3 What are viruses? 246
  (Inquiry) **Lab** Bacterial Growth and Disinfectants 254

Chapter 8 **Protists and Fungi** 262
Lesson 1 What are protists?...264
Lesson 2 What are fungi? 276
  (Inquiry) **Lab** What does a lichen look like?...............................286

Chapter 9 **Plant Diversity** 294
Lesson 1 What is a plant?...296
Lesson 2 Seedless Plants 306
  (Inquiry) **Skill Practice** How do differences in plant structures reflect their
    environments? 311
Lesson 3 Seed Plants 312
  (Inquiry) **Lab** Compare and Contrast Extreme Plants...........................322

Chapter 10 **Plant Processes and Reproduction**.......................330
Lesson 1 Energy Processing in Plants 332
Lesson 2 Plant Responses 340
  (Inquiry) **Skill Practice** What happens to seeds if you change the intensity
    of light?..349
Lesson 3 Plant Reproduction 350
  (Inquiry) **Lab** Design a Stimulating Environment for Plants 360

TABLE OF CONTENTS

Table of Contents

Student Resources

Science Skill Handbook **SR-2**
  Scientific Methods SR-2
  Safety Symbols SR-11
  Safety in the Science Laboratory SR-12

Math Skill Handbook **SR-14**
  Math Review SR-14
  Science Application SR-24

Foldables Handbook **SR-29**

Reference Handbook **SR-40**
  Periodic Table of the Elements SR-40
  Diversity of Life: Classification of Living Organisms........................ SR-42
  Use and Care of a Microscope SR-46

Glossary **G-2**

Index **I-2**

Credits **C-2**

TABLE OF CONTENTS

Inquiry

Inquiry Launch Labs

7-1 How small are bacteria? 231
7-2 How do bacteria affect the environment? 239
7-3 How quickly do viruses replicate? 247
8-1 How does a protist react to its environment? 265
8-2 Is there a fungus among us? 277
9-1 What is a plant? 297
9-2 Which holds more water? 307
9-3 What characteristics do seeds have in common? 313
10-1 How can you show the movement of materials inside a plant? ... 333
10-2 How do plants respond to stimuli? 341
10-3 How can you identify fruits? 351

Inquiry MiniLabs

7-1 How does a slime layer work? 233
7-2 Can decomposition happen without oxygen? 240
7-3 How do antibodies work? 252
8-1 How can you model the movement of an amoeba? 271
8-2 What do fungal spores look like? 280
9-1 How does water loss from a leaf relate to the thickness of the cuticle? ... 299
9-3 How can you determine the stems, roots, and leaves of plants? ... 316
10-1 Can you observe plant processes? 336
10-2 When will plants flower? 345
10-3 Can you model a flower? 356

Inquiry **Skill Practice**

7-2 How do lab techniques affect an investigation? 245

9-2 How do differences in plant structures reflect their environments?........................ 311

10-2 What happens to seeds if you change the intensity of light? 349

Inquiry **Labs**

7-3 Bacterial Growth and Disinfectants.. 254

8-2 What does a lichen look like? 286

9-3 Compare and Contrast Extreme Plants 322

10-3 Design a Stimulating Environment for Plants 360

Features

GREEN
SCIENCE

8-1 The Benefits of Algae 275

10-1 Deforestation and Carbon Dioxide in the Atmosphere...................................... 339

Unit 2

FROM BACTERIA TO PLANTS

1700

1753
Swedish botanist Carl Linnaeus publishes *Species Plantarum,* a list of all plants known to him and the starting point of modern plant nomenclature.

1800

1892
Russian botanist Dmitri Iwanowski discovers the first virus while studying tobacco mosaic disease. Iwanowski finds that the cause of the disease is small enough to pass through a filter made to trap all bacteria.

1900

1930s
The development of commercial hybrid crops begins in the United States.

1983
Luc Montagnier's team at the Pasteur Institute in France isolates the retrovirus now called HIV.

Nature of SCIENCE

Technology

Some people wonder why governments have invested so much money to explore space. Why not solve problems here on Earth using that money? What did all of that money buy?

Technology is the practical application of science to commerce or industry. Science and technology depend on each other. Once scientists understand a scientific concept, they apply the science to new technologies. Today, many technologies originally developed for space are solving problems for people worldwide. For example, sensors designed to remotely measure the temperature of distant stars led to the development of the thermometer shown in **Figure 1.** When pointed toward the ear canal, the thermometer provides an accurate temperature reading in 2 seconds. The images in **Figure 2** show other technologies developed for space that are now used on Earth.

Figure 1 A technology originally developed to measure the temperature of stars now enables a parent to easily and quickly determine whether a child has a fever.

Figure 2 Medical professionals use technologies originally developed for space to help improve the health of patients.

To monitor the health of astronauts during space walks, space suits contain tiny sensors that measure astronauts' temperature, respiration, and cardiac activity. The technology that led to the development of these sensors is now used on Earth to monitor people's health. ▶

◀ Hospitals can monitor patients from a central nurses' station using electronics similar to those used in space suits. Sensors in infants' clothes can monitor a baby's breathing while the baby sleeps.

Doctors use sensors developed to study the effect of weightlessness on the muscles of astronauts to monitor repeated muscle movements that can lead to carpal tunnel syndrome in the workplace. ▶

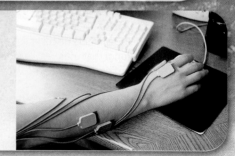

Space technology improves health worldwide.

Worldwide, people contribute to and benefit from space technologies. For example, a Canadian company produced robotic arms for the space shuttle. They collaborated with a medical school to develop instruments that enable surgeons to perform microscopic surgery on the brain. Chinese scientists developed a high-resolution X-ray imaging system for spacecraft. These X-rays are safer, faster, and more accurate than previous X-rays. Now, doctors can use the system to more accurately diagnose diseases. In addition, a Spanish company developed a navigation system for the blind based on space navigation technology.

Science and technology cannot solve all problems. But together, they help medical professionals keep people healthy and improve the quality of life for everyone on Earth.

Inquiry MiniLab
25 minutes

What can you invent from space technology?

Space technology may have more uses than are currently known. Your job is to think of a new use for a joystick.

1. Work with a partner to discuss ways in which technology used in space might be used to improve health. Consider various types of illnesses or medical conditions in which a person's abilities to do everyday tasks might be helped by space technology.

2. Select one idea and develop it. Draw pictures of your device to show how it will work.

Analyze and Conclude

1. **Explain** How does your invention improve the ability of medical professionals or the lives of people with medical conditions?

2. 🔑 **Key Concept** Why is your device a space technology?

Surgeons use joysticks, similar to those used to control the Lunar Rover, to perform surgery on a patient who is thousands of miles away.

The temperature inside an astronaut's space suit can become extremely hot. So NASA developed a technology that circulates a cool fluid through tubes built into the suit.

Scientists applied this technology and developed a therapy for people with multiple sclerosis (MS). MS is a disease that slows the transfer of nerve signals from the brain. MS can affect the ability to think, speak, and control movement. Studies show that a slight decrease in body temperature can restore the transfer of some nerves signals. Therefore, scientists developed cooling suits for patients with MS based on NASA's cooling space suits.

Chapter 7

Bacteria and Viruses

THE BIG IDEA

What are bacteria and viruses and why are they important?

Color-enhanced SEM Magnification: 560×

Inquiry Are robots attacking?

You might think this photo shows robots landing on another planet. Actually, this is a picture of viruses attacking a type of unicellular organism called a bacterium (plural, bacteria). Many viruses can attach to the surface of one bacterium.

- Do you think the bacterium is harmful? Are the viruses?

- What do you think happens after the viruses attach to the bacterium?

- What are viruses and bacteria and why are they important?

Get Ready to Read

What do you think?

Before you read, decide if you agree or disagree with each of these statements. As you read this chapter, see if you change your mind about any of the statements.

1 A bacterium does not have a nucleus.

2 Bacteria cannot move.

3 All bacteria cause diseases.

4 Bacteria are important for making many types of food.

5 Viruses are the smallest living organisms.

6 Viruses can replicate only inside an organism.

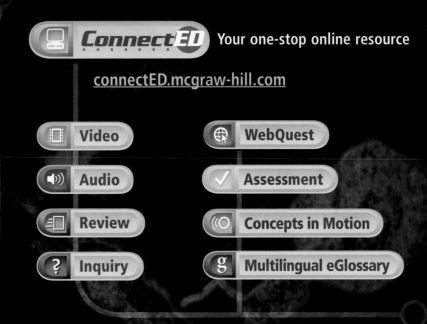

ConnectED Your one-stop online resource

connectED.mcgraw-hill.com

- Video
- Audio
- Review
- Inquiry
- WebQuest
- Assessment
- Concepts in Motion
- Multilingual eGlossary

Reading Guide

Key Concepts
ESSENTIAL QUESTIONS

- What are bacteria?

Vocabulary

bacterium p. 231
flagellum p. 234
fission p. 234
conjugation p. 234
endospore p. 235

 Multilingual eGlossary

Video **BrainPOP®**

What are bacteria?

Color-enhanced SEM Magnification: 560×

nquiry How clean is this surface?

This photo shows a microscopic view of the point of a needle. The small orange things are bacteria. Bacteria are everywhere, even on surfaces that appear clean. Do you think bacteria are living or nonliving?

How small are bacteria?

Bacteria are tiny cells that can be difficult to see, even with a microscope. You might be surprised to learn that bacteria are found all around you, including in the air, on your skin, and in your body. One way of understanding how small bacteria are is to model their size.

1 Read and complete a lab safety form.

2 Examine the size of a **baseball** and a **2.5-gal. bucket.** Estimate how many baseballs you think would fit inside the bucket.

3 As a class, count how many baseballs it takes to fill the bucket.

Think About This

1. How much larger is the bucket than a baseball?

2. If your skin cells were the size of the bucket and bacteria were the size of the baseballs, how many bacterial cells would fit on a skin cell?

3. **Key Concept** Why do you think you cannot see bacteria on your skin or on your desk?

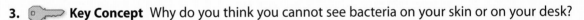

Characteristics of Bacteria

Did you know that billions of tiny organisms too small to be seen surround you? These organisms, called bacteria, even live inside your body. **Bacteria** (singular, bacterium) *are microscopic prokaryotes.* You might recall that a prokaryote is a unicellular organism that does not have a nucleus or other membrane-bound organelles.

Bacteria live in almost every habitat on Earth, including the air, glaciers, the ocean floor, and in soil. A teaspoon of soil can contain between 100 million and 1 billion bacteria. Bacteria also live in or on almost every organism, both living and dead. Hundreds of species of bacteria live on your skin. In fact, your body contains more bacterial cells than human cells! The bacteria in your body outnumber human cells by 10 to 1.

Key Concept Check What are bacteria?

Other prokaryotes, called archaea (ar KEE uh; singular, archaean), are similar to bacteria and share many characteristics with them, including the lack of membrane-bound organelles. Archaea can live in places where few other organisms can survive, such as very warm areas or those with little oxygen. Both bacteria and archaea are important to life on Earth.

WORD ORIGIN

bacteria
from Greek *bakterion*, means "small staff"

Make a folded book from a sheet of notebook paper. Label it as shown. Use your book to organize your notes on the characteristics of bacteria.

Characteristics of Bacteria

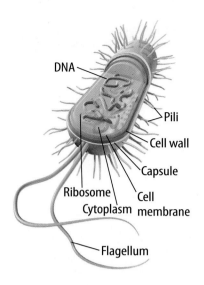

DNA
Pili
Cell wall
Capsule
Ribosome
Cytoplasm
Cell membrane
Flagellum

▲ **Figure 1** Bacteria have a cell membrane and contain cytoplasm.

Structure of Bacteria

A typical bacterium, such as the one shown in **Figure 1,** consists of cytoplasm and DNA surrounded by a cell membrane and a cell wall. The cytoplasm also contains ribosomes. Most bacteria have DNA that is one coiled, circular chromosome. Many bacteria also have one or more small circular pieces of DNA called plasmids that are separate from its other DNA.

Some bacteria have specialized structures that help them survive. For example, the bacterium that causes pneumonia (noo MOH nyuh), an inflammation of the lungs, has a thick covering, or capsule, around its cell wall. The capsule protects the bacterium from drying out. It also prevents white blood cells from surrounding and antibiotics from entering it. Many bacteria have capsules with hairlike structures called pili (PI li) that help the bacteria stick to surfaces.

Size and Shapes of Bacteria

Bacteria are much smaller than plant or animal cells. Bacteria are generally only 1–5 micrometers (µm) (1 m = 1 million µm) wide, while an average eukaryotic cell is 10–100 µm wide. Scientists estimate that as many as 100 bacteria could be lined up across the head of a pin. As shown in **Figure 2,** bacteria generally have one of three basic shapes.

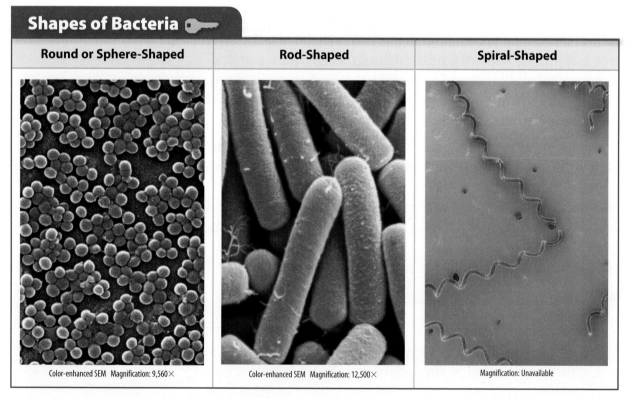

Shapes of Bacteria 🔑

Round or Sphere-Shaped	Rod-Shaped	Spiral-Shaped
Color-enhanced SEM Magnification: 9,560×	Color-enhanced SEM Magnification: 12,500×	Magnification: Unavailable

Figure 2 Bacteria are generally shaped like a sphere, a rod, or a spiral.

✓ **Visual Check** What are the three basic shapes of bacteria?

How does a slime layer work?

Bacteria have a gelatinlike, protective coating called a slime layer on the outside of their cell walls. A slime layer can help a bacterium attach to surfaces or reduce water loss.

1. Read and complete a lab safety form.
2. Cut two 2-cm-wide strips from the long side of a **synthetic kitchen sponge**.
3. Soak both strips in **water**. Remove them from the water and squeeze out the excess water. Both strips should be damp.
4. Completely coat one strip with **hair-styling gel** to simulate a slime layer.
5. Place both strips on a **plate** and let them sit overnight.

Analyze and Conclude

1. **Describe** the appearance of the two strips in your Science Journal. How do they differ?

2. **Key Concept** Explain how a slime layer might be beneficial to a bacterium when moving or finding food.

Obtaining Food and Energy

Bacteria live in many places. Because these environments are very different, bacteria obtain food in various ways. Some bacteria take in food and break it down and obtain energy. Many of these bacteria feed on dead organisms or organic waste, as shown in **Figure 3.** Others take in their nutrients from living hosts. For example, bacteria that cause tooth decay live in dental plaque on teeth and feed on sugars in the foods you eat and the beverages you drink.

Some bacteria make their own food. These bacteria use light energy and make food, like most plants do. These bacteria live where there is a lot of light, such as the surface of lakes and streams. Other bacteria use energy from chemical reactions and make their food. These bacteria live in places where there is no sunlight, such as the dark ocean floor.

Figure 3 This banana is rotting because bacteria are breaking it down to use it for food.

Key Concept Check How do bacteria obtain food?

Most organisms, including humans, cannot survive without oxygen. However, certain bacteria do not need oxygen to survive. These bacteria are called anaerobic (a nuh ROH bihk) bacteria. Bacteria that need oxygen are called aerobic (er OH bihk) bacteria. Most bacteria in the environment are aerobic.

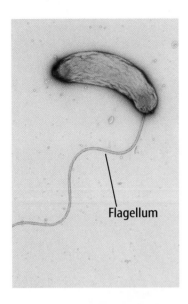

▲ **Figure 4** Some bacteria move using a flagellum.

Movement

Some bacteria are able to move around to find the resources that they need to survive. These bacteria have special structures for movement. *Many bacteria have long whiplike structures called* **flagella** (fluh JEH luh; singular, flagellum), as shown in **Figure 4.** Others twist or spiral as they move. Still other bacteria use their pili like grappling hooks or make threadlike structures that enable them to push away from a surface.

Reproduction

You might recall that organisms reproduce asexually or sexually. Bacteria reproduce asexually by fission. **Fission** *is cell division that forms two genetically identical cells.* Fission can occur quickly—as often as every 20 minutes under ideal conditions.

Bacteria produced by fission are identical to the parent cell. However, genetic variation can be increased by a process called conjugation, shown in **Figure 5.** *During* **conjugation** (kahn juh GAY shun), *two bacteria of the same species attach to each other and combine their genetic material.* DNA is transferred between the bacteria. This results in new combinations of genes, increasing genetic diversity. New organisms are not produced during conjugation, so the process is not considered reproduction.

Reading Check How does conjugation increase the genetic diversity of bacteria?

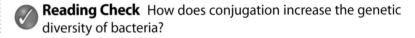

Conjugation

Review   Personal Tutor

Figure 5 Conjugation results in genetic diversity by transferring DNA between two bacteria cells.

Visual Check What structure does the donor cell use to connect to the recipient cell?

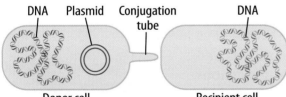

DNA   Plasmid   Conjugation tube   DNA

Donor cell   Recipient cell

❶ The donor cell and recipient cell both have circular chromosomal DNA. The donor cell also has DNA as a plasmid. The donor cell forms a conjugation tube and connects to the recipient cell.

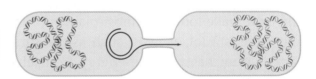

❷ The conjugation tube connects both cells. The plasmid splits in two and one plasmid strand moves through the conjugation tube into the recipient cell.

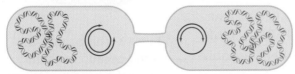

❸ The complimentary strands of the plasmids are completed in both bacteria.

❹ With the new plasmids complete, the bacteria separate from each other. The recipient cell now contains plasmid DNA from the donor cell as well as its own chromosomal DNA.

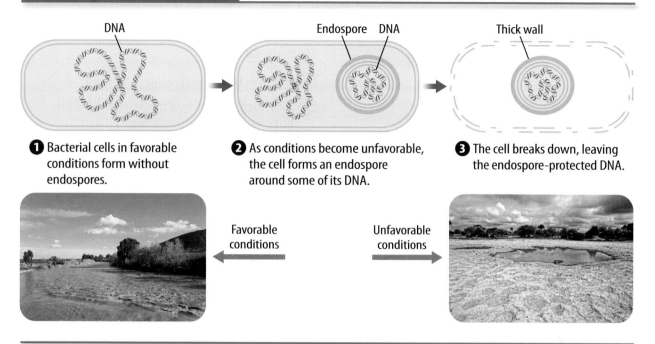

1 Bacterial cells in favorable conditions form without endospores.

2 As conditions become unfavorable, the cell forms an endospore around some of its DNA.

3 The cell breaks down, leaving the endospore-protected DNA.

Favorable conditions ← | → Unfavorable conditions

Figure 6 An endospore protects a bacterium.

Endospores

Sometimes environmental conditions are unfavorable for the survival of bacteria. In these cases, some bacteria can form endospores. *An* **endospore** (EN doh spor) *forms when a bacterium builds a thick internal wall around its chromosome and part of the cytoplasm,* as shown in **Figure 6.** An endospore can protect a bacterium from intense heat, cold, or drought. It also enables a bacterium to remain dormant for months or even centuries. The ability to form endospores enables bacteria to survive extreme conditions that would normally kill them.

Archaea

Prokaryotes called archaea were once considered bacteria. Like a bacterium, an archaean has a cell wall and no nucleus or membrane-bound organelles. Its chromosome is also circular, like those in bacteria. However, there are some important differences between archaea and bacteria. The ribosomes of archaea more closely resemble the ribosomes of eukaryotes than those of bacteria. Archaea also contain molecules in their plasma membranes that are not found in any other known organisms. Archaea often live in extreme environments, such as hot springs and salt lakes. Some scientists refer to archaea as extremophiles (ik STREE muh filez)—a term that means "those that love extremes."

Math Skills ✕÷

Use a Formula

Each time bacteria undergo fission, the population doubles. Use an equation to calculate how many bacteria there are: $n = x \times 2^f$ where n is the final number of bacteria, x is the starting number of bacteria, and f is the number of times that fission occurs.

Example: 100 bacteria undergo fission 3 times.

$f = 3$, so 2^f is 2 multiplied by itself 3 times. ($2 \times 2 \times 2 = 8$)

$n = 100 \times 8 = 800$ bacteria

Practice

How many bacteria would there be if 1 bacterium underwent fission 10 times?

📑 **Review**

• **Math Practice**
• **Personal Tutor**

Lesson 1 Review

Visual Summary

Bacteria are unicellular prokaryotes.

Many bacteria feed on dead organic matter.

Bacteria can increase genetic diversity by sharing DNA through conjugation.

FOLDABLES

Use your lesson Foldable to review the lesson. Save your Foldable for the project at the end of the chapter.

What do you think NOW?

You first read the statements below at the beginning of the chapter.

1. A bacterium does not have a nucleus.

2. Bacteria cannot move.

Did you change your mind about whether you agree or disagree with the statements? Rewrite any false statements to make them true.

Use Vocabulary

1. **Use the term** *bacteria* in a sentence.

2. The long whiplike structure that some bacteria use for movement is a(n) _____.

3. **Define** *conjugation* in your own words.

Understand Key Concepts

4. **Describe** a typical bacterium.

5. Which is NOT a common bacteria shape?
   - **A.** rod
   - **B.** sphere
   - **C.** spiral
   - **D.** square

6. **Contrast** fission and conjugation.

Interpret Graphics

7. **Identify** Copy and complete the table below to identify shapes of bacteria.

Bacterial Shapes	Illustration

Critical Thinking

8. **Describe** how a bacterium's small size could be an advantage or a disadvantage for its survival.

9. **Explain** how bacteria might find food and survive in an environment where few other organisms live.

10. **Analyze** how bacteria that can form endospores would have an advantage over bacteria that cannot form endospores.

Math Skills ✕ ÷ +

Review
Math Practice

11. How many bacteria would there be if fission occurred 4 times with 1,000 bacteria?

Cooking Bacteria!

How Your Body Is Like Bleach

When it comes to killing germs, few things work as well as household bleach. How does bleach kill bacteria? Believe it or not, killing bacteria with bleach and boiling an egg involve similar processes.

▼ After cooking, egg proteins become a tangled mass.

Eggs are made mostly of proteins. Proteins are complex molecules in all plant and animal tissues. Proteins have specific functions that are dependent on the protein's shape. A protein's function changes if its shape is changed. When you cook an egg, the thermal energy transferred to the egg causes changes to the shape of the egg's proteins. Think of the firm texture of a cooked egg. When the egg's proteins are heated, they become a tangled mass.

▲ Before cooking, the proteins in eggs remain unfolded and change shape easily.

▼ Bacteria also contain proteins that change shape when exposed to heat.

A common ingredient in bleach is also found in your body's immune cells. ▶

Like eggs, bacteria also contain proteins. When bacteria are exposed to high temperatures, their proteins change shape, similar to those in a boiled egg. But what is the connection with bleach? Scientists have discovered that an ingredient in bleach, hypochlorite (hi puh KLOR ite), also causes proteins to change shape. The bacterial proteins that are affected by bleach are needed for the bacteria's growth. When the shape of those proteins changes, they no longer function properly, and the bacteria die.

Scientists also know now that your body's immune cells produce hypochlorite. Your body protects itself with the same chemical you can use to clean your kitchen!

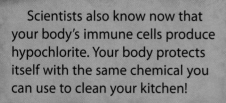

It's Your Turn

RESEARCH AND REPORT A bacterial infection often causes inflammation, or a response to tissue damage that can include swelling and pain. Research and report on what causes inflammation.

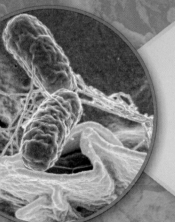

Bacteria in Nature

Reading Guide

Key Concepts
ESSENTIAL QUESTIONS

- How can bacteria affect the environment?
- How can bacteria affect health?

Vocabulary

decomposition p. 240
nitrogen fixation p. 240
bioremediation p. 241
pathogen p. 242
antibiotic p. 242
pasteurization p. 243

 Multilingual eGlossary

Video

What's Science Got to do With It?

nquiry Why does this larva glow?

Some bacteria have the ability to glow in the dark. The moth larva shown on this page is filled with many such bacteria. These bacteria produce toxins that can slowly kill the animal. A chemical reaction within each bacterium makes the larva's body appear to glow.

How do bacteria affect the environment?

Bacteria are everywhere in your environment. They are in the water, in the air, and even in some foods.

1. Read and complete a lab safety form.
2. Carefully examine the contents of the two **bottles** provided by your teacher.
3. Record your observations in your Science Journal.

Think About This

1. Compare your observations of bottle A to those of bottle B. Which one appears to have more bacteria in it? Support your answer.

2. 🔑 **Key Concept** Based on your observations, how could bacteria affect the environment around you?

Beneficial Bacteria

When you hear about bacteria, you probably think about getting sick. However, only a fraction of all bacteria cause diseases. Most bacteria are beneficial. In fact, many organisms, including humans, depend on bacteria to survive. Some types of bacteria help with digestion and other body processes. For example, one type of bacteria in your intestines makes vitamin K, which helps your blood clot properly. Several others help break down food into smaller particles. Another type of bacteria called *Lactobacillus* lives in your intestines and prevents harmful bacteria from growing.

Animals benefit from bacteria as well. Without bacteria, some organisms, such as the cow pictured in **Figure 7,** wouldn't be able to digest the plants they eat. Bacteria and other microscopic organisms live in a large section of the cow's stomach called the rumen. The bacteria help break down a substance in grass called cellulose into smaller molecules that the cow can use.

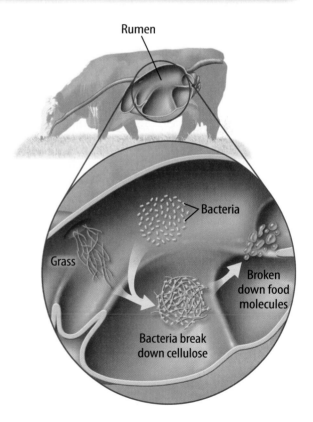

Figure 7 Cows get help digesting the cellulose in plants from the bacteria that live in their rumen—one of four stomach sections.

✅ **Visual Check** What role do bacteria play in a cow's digestion?

Decomposition

What do you think would happen if organic waste such as food scraps and dead leaves never decayed? **Decomposition,** *the breaking down of dead organisms and organic waste,* is an important process in nature. When a tree dies, bacteria and other decomposing organisms feed on the dead organic matter. As decomposers break down the tree, they release molecules such as carbon and phosphorus into the soil that other organisms can then take in and use for life processes.

Nitrogen Fixation

Organisms use nitrogen to make proteins. Although about 78 percent of the atmosphere is nitrogen gas, it is in a form that plants and animals cannot use. Some plants can obtain nitrogen from bacteria. These plants have special structures called nodules, shown in **Figure 8,** on their roots. Bacteria in the nodules convert nitrogen from the atmosphere into a form usable to plants. **Nitrogen fixation** *is the conversion of atmospheric nitrogen into nitrogen compounds that are usable by living things.*

Key Concept Check What are some ways that bacteria are beneficial to the environment?

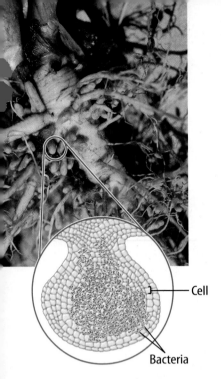

 — Cell

Bacteria

Figure 8 The roots of some plants have nodules that contain nitrogen-fixing bacteria.

Inquiry MiniLab — **20 minutes**

Can decomposition happen without oxygen?

You have just read that bacteria play an important role as decomposers in the environment. How do you think decomposition differs in aerobic and anaerobic environments?

1. Read and complete a lab safety form.

2. Obtain two **self-sealing plastic bags** from your teacher. Use a **permanent marker** and label one bag *Bag A* and the other *Bag B.*

3. Place a **slice of apple** in bag A. Seal the bag leaving as much air as possible inside of it. Set the bag aside.

4. Place **another slice of apple** in bag B. Carefully squeeze the bag to remove as much air as possible before sealing it. Place both bags in the location specified by your teacher and leave overnight.

5. The next lab day, observe both bags. Note the appearance of the apples. Record your observations in your Science Journal.

6. Carefully dispose of both bags according to your teacher's directions.

Analyze and Conclude

1. **Determine** which apple changed the most. How could you tell? List specific evidence to support your answer.

2. **Draw Conclusions** Does decomposition occur faster, slower, or not at all in environments without oxygen? Justify your answer.

3. **Key Concept** Summarize why bacteria are considered important decomposers.

Bioremediation

Can you imagine an organism that eats pollution? Some bacteria do just that. *The use of organisms, such as bacteria, to clean up environmental pollution is called* **bioremediation** (bi oh rih mee dee AY shun). These organisms often break down harmful substances, such as sewage, into less harmful material that can be used as landfill or fertilizers.

Bacteria are commonly used to clean up areas that have been contaminated by oil or harmful plastics. Some kinds of bacteria can even help clean up radioactive waste, such as uranium in the abandoned mine fields shown in **Figure 9.** In many cases, without using bacteria, the substances would take centuries to break down and would contaminate soils and water.

 Reading Check Why might using bacteria to clean up environmental spills be a good option?

Figure 9 These bacteria clean the environment by removing harmful uranium from the water.

Bacteria and Food

Would you like a side of bacteria with that sandwich? If you have eaten a pickle lately, you might have had some. Some pickles are made when the sugar in cucumbers is converted into an acid by a specific type of bacteria. Pickles are just one of the many food products made with the help of bacteria. Bacteria are used to make foods such as yogurt, cheese, buttermilk, vinegar, and soy sauce. Bacteria are even used in the production of chocolate. They help break down the covering of the cocoa bean during the process of making chocolate. Bacteria are responsible for giving chocolate some of its flavor.

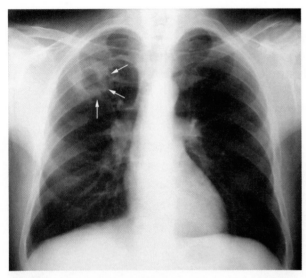

Figure 10 In an X-ray, the lungs of a person with tuberculosis may show pockets or scars where bacterial infection has begun.

Visual Check How do you think the bacteria that made this person sick entered his or her body?

Harmful Bacteria

Of the 5,000 known species of bacteria, relatively few are considered **pathogens** (PA thuh junz)—*agents that cause disease.* Some pathogens normally live in your body, but cause illness only when your immune system is weakened. For example, the bacterium *Streptococcus pneumoniae* lives in the throats of most healthy people. However, it can cause pneumonia if a person's immune system is weakened. Other bacterial pathogens can enter your body through a cut, the air you breathe, or the food you eat. Once inside your body, they can reproduce and cause disease.

 Key Concept Check Describe one way that bacteria can be harmful to health.

Bacterial Diseases

Bacteria can harm your body and cause disease in one of two ways. Some bacteria make you sick by damaging tissue. For example, the disease tuberculosis, shown in **Figure 10,** is caused by a bacterium that invades lung tissue and breaks it down for food. Other bacteria cause illness by releasing toxins. For example, the bacterium *Clostridium botulinum* can grow in improperly canned foods and produce toxins. If the contaminated food is eaten, the toxins can cause food poisoning, resulting in paralyzed limbs or even death.

Treating Bacterial Diseases Most bacterial diseases in humans can be treated with antibiotics. **Antibiotics** (an ti bi AH tihks) *are medicines that stop the growth and reproduction of bacteria.* Many antibiotics work by preventing bacteria from building cell walls. Others affect ribosomes in bacteria, interrupting the production of proteins.

Many types of bacteria have become **resistant** to antibiotics over time. Some diseases, such as tuberculosis, pneumonia, and meningitis, are now more difficult to treat.

Bacterial Resistance How do you think bacteria become resistant to antibiotics? This process, shown in **Figure 11,** can happen over a long or short period of time depending on how quickly the bacteria reproduce. Random mutations occur to a bacterium's DNA that enable it to survive or "resist" a specific antibiotic. If that antibiotic is used as a treatment, only the bacteria with the mutation will survive.

Over time, the resistant bacteria will reproduce and become more common. The antibiotic is no longer effective against that bacterium, and a different antibiotic must be used to fight the disease. Scientists are always working to develop more effective antibiotics to which bacteria have not developed resistances.

 Reading Check How do bacteria develop resistance to antibiotics?

Food Poisoning

All food, unless it has been treated or processed, contains bacteria. Over time these bacteria reproduce and begin breaking down the food, causing it to spoil. As you read on the previous page, eating food contaminated by some bacteria can cause food poisoning. By properly treating or processing food and killing bacteria before the food is stored or eaten, it is easier to avoid food poisoning and other illnesses.

Pasteurization (pas chuh ruh ZAY shun) *is a process of heating food to a temperature that kills most harmful bacteria.* Products such as milk, ice cream, yogurt, and fruit juice are usually pasteurized in factories before they are transported to grocery stores and sold to you. After pasteurization, foods are much safer to eat. Foods do not spoil as quickly once they have been pasteurized. Because of pasteurization, food poisoning is much less common today than it was in the past.

 Key Concept Check How does pasteurization affect human health?

Figure 11 A population of bacteria can develop resistance to antibiotics after being exposed to them over time.

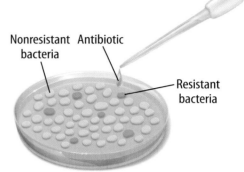

Nonresistant bacteria   Antibiotic   Resistant bacteria

1 An antibiotic is added to a colony of bacteria. A few of the bacteria have mutations that enable them to resist the antibiotic.

2 The antibiotic kills most of the nonresistant bacteria. The resistant bacteria survive and reproduce, creating a growing colony of bacteria.

3 Surviving bacteria are added to another plate containing more of the same antibiotic.

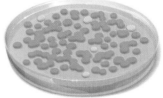

4 The antibiotic now affects only a small percentage of the bacteria. The surviving bacteria continue to reproduce. Most of the bacteria are resistant to the antibiotic.

Lesson 2 Review

Visual Summary

Bacteria can help some organisms, including humans and cows, digest food.

Bacteria can be used to remove harmful substances such as uranium.

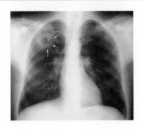

Some bacteria are pathogens, and cause diseases in humans and other organisms.

FOLDABLES

Use your lesson Foldable to review the lesson. Save your Foldable for the project at the end of the chapter.

What do you think NOW?

You first read the statements below at the beginning of the chapter.

3. All bacteria cause diseases.

4. Bacteria are important for making many types of food.

Did you change your mind about whether you agree or disagree with the statements? Rewrite any false statements to make them true.

Use Vocabulary

1 **Distinguish** between an antibiotic and a pathogen.

2 **Define** *bioremediation* using your own words.

3 **Use the term** *pasteurization* in a sentence.

Understand Key Concepts

4 Which of the following is NOT a beneficial use of bacteria?

A. bioremediation   C. food poisoning
B. decomposition   D. nitrogen fixation

5 **Compare** the benefits of nitrogen fixation and decomposition.

6 **Analyze** the importance of bacteria in food production.

Interpret Graphics

7 **Examine** the figure below and describe what would happen if bacteria were not present.

8 **Identify** Copy and complete the graphic organizer below to identify ways that bacteria can be beneficial.

Beneficial Bacteria

Critical Thinking

9 **Evaluate** the effect of all bacteria becoming resistant to antibiotics.

How do lab techniques affect an investigation?

Materials

petri dish

jar with samples

forceps

dissecting microscope

black light

Safety

Pathogens such as bacteria cover almost every surface. When you touch a surface, you transfer particles from that surface to your skin and then to other objects you touch. Your teacher has spread a substance that simulates bacteria on some surfaces in this lab. You will be divided into two groups. Each group will perform the same lab activity but will use slightly different laboratory techniques.

Learn It

In a laboratory it is important to be very careful to keep surfaces as free from contamination as possible. Scientists follow specific **lab techniques** very carefully to prevent contamination that could affect results.

Try It

1. Read and complete a lab safety form.

2. Put on a pair of gloves. Select a Petri dish from the stack. Open the Petri dish and follow the directions on the slip of paper.

3. Go to the station with the jar. Open the jar and use forceps to remove an item. Place the item in your Petri dish. Close the jar. Follow the directions again.

4. Take your Petri dish to the dissecting microscope and examine your object. Sketch the object in your Science Journal.

5. Observe the surfaces in your work area as your teacher shines a black light over them.

Apply It

6. What surfaces light up the most under the black light?

7. What do you see when you use the black light?

8. 🔑 **Key Concept** What difference do you see in the lab areas used by the two groups? Based on your observations, how do you think this difference affects which techniques are used in labs and hospitals?

Reading Guide

Key Concepts
ESSENTIAL QUESTIONS

- What are viruses?
- How do viruses affect human health?

Vocabulary

virus p. 247

antibody p. 251

vaccine p. 252

g Multilingual eGlossary

What are viruses?

Inquiry Painted Flowers?

The streaking patterns on the petals of these tulips are not painted on but are caused by a virus. Tulips with these patterns are prized for their beautiful appearance. How do you think a virus could cause this flower's pattern? Do you think all viruses are harmful?

How quickly do viruses replicate?

One characteristic that viruses share is the ability to produce many new viruses from just one virus. In this lab you can use grains of rice to model virus replication. Each grain of rice represents one virus.

Generation	First	Second	Third
Number of "viruses"			

1 Read and complete a lab safety form.

2 Copy the table above into your Science Journal.

3 Estimate the number of **grains of rice** in the **fishbowl** and record this number for the first generation.

4 One student will add the contents of his or her **cup** to the fishbowl. Estimate how many viruses are now in the fishbowl and record your estimate for the second generation.

5 The rest of the class will add the contents of their cups to the fishbowl. Estimate the number of viruses and record that number of viruses for the third generation.

Think About This

1. Recall that bacteria double every generation. How does the number of viruses produced in each generation compare with the number of bacteria produced in each generation?

2. 🔑 **Key Concept** How could the rate at which viruses are produced affect human health?

Characteristics of Viruses

Do chicken pox, mumps, measles, and polio sound familiar? You might have received shots to protect you from these diseases. You might have also received a shot to protect you from influenza, commonly known as the flu. What do these diseases have in common? They are caused by different viruses. *A* **virus** *is a strand of DNA or RNA surrounded by a layer of protein that can infect and replicate in a host cell.* If you have had a cold, you have been infected by a virus.

A virus does not have a cell wall, a nucleus, or any other organelles present in cells. The smallest viruses are between 20 and 100 times smaller than most bacteria. Recall that about 100 bacteria would fit across the head of a pin. Viruses can have different shapes, such as the crystal, cylinder, sphere, and bacteriophage (bak TIHR ee uh fayj) shapes shown in **Figure 12.**

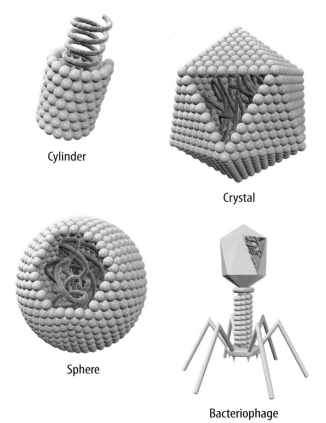

Cylinder

Crystal

Sphere

Bacteriophage

Figure 12 Viruses have a variety of shapes.

Make a folded book from a sheet of paper. Label it as shown. Use it to organize your notes on the replication sequence of a virus.

Viral Replication

Dead or Alive?

Do you think that viruses are living things? Scientists do not consider viruses to be alive because they do not have all the characteristics of a living organism. Recall that living things are organized, respond to stimuli, use energy, grow, and reproduce. Viruses cannot do any of these things. A virus can make copies of itself in a process called replication, but it must rely on a living organism to do so.

Key Concept Check Are viruses alive? Explain why or why not.

Viruses and Organisms

Viruses must use organisms to carry on the processes that we usually associate with a living cell. Viruses have no organelles so they are not able to take in nutrients or use energy. They also cannot replicate without using the cellular parts of an organism. Viruses must be inside a cell to replicate. The living cell that a virus infects is called a host cell.

When a virus enters a cell, as shown in **Figure 13,** it can either be active or latent. Latent viruses go through an inactive stage. Their genetic material becomes part of the host cell's genetic material. For a period of time, the virus does not take over the cell to produce more viruses. In some cases, viruses have been known to be inactive for years and years. However, once it becomes active, a virus takes control of the host cell and replicates.

Figure 13 A virus infects a cell by inserting its DNA or RNA into the host cell. It then directs the host cell to make new viruses.

Visual Check What occurs when a virus becomes latent?

Viral Replication 🔑

Concepts in Motion Animation

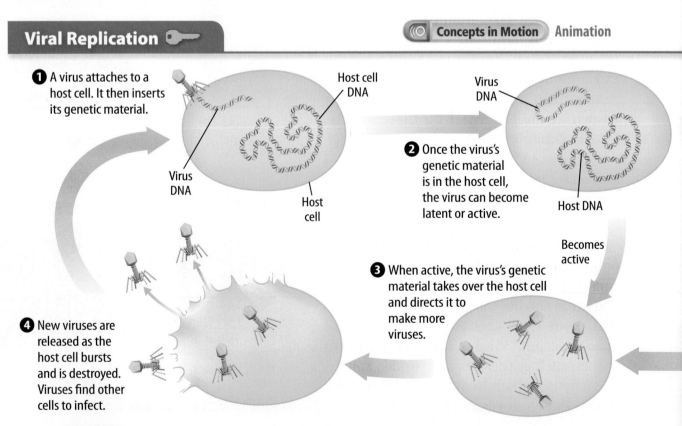

❶ A virus attaches to a host cell. It then inserts its genetic material.

Host cell DNA

Virus DNA

Virus DNA

Host cell

Host DNA

❷ Once the virus's genetic material is in the host cell, the virus can become latent or active.

Becomes active

❸ When active, the virus's genetic material takes over the host cell and directs it to make more viruses.

❹ New viruses are released as the host cell bursts and is destroyed. Viruses find other cells to infect.

Replication

As you read earlier, a virus can make copies of itself in a process called replication, shown in **Figure 13.** A virus cannot infect every cell. A virus can only attach to a host cell with specific molecules on its cell wall or cell membrane. These molecules enable the virus to attach to the host cell. This is similar to the way that only certain electrical plugs can fit into an outlet on a wall. After a virus attaches to the host cell, its DNA or RNA enters the host cell. Once inside, the virus either starts to replicate or becomes latent, also shown in **Figure 13.** After a virus becomes active and replicates in a host cell, it destroys the host cell. Copies of the virus are then released into the host organism, where they can infect other cells.

Mutations

As viruses replicate, their DNA or RNA frequently mutates, or changes. These **mutations** enable viruses to adjust to changes in their host cells. For example, the molecules on the outside of host cells change over time to prevent viruses from attaching to the cell. As viruses mutate, they are able to produce new ways to attach to host cells. These changes happen so rapidly that it can be difficult to cure or prevent viral diseases before they mutate again.

✓ **Reading Check** How does mutation enable viruses to continue causing disease?

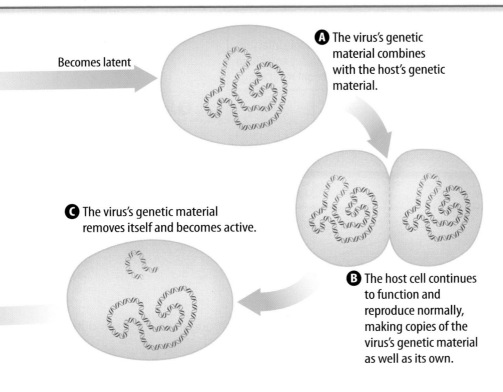

Becomes latent

A The virus's genetic material combines with the host's genetic material.

C The virus's genetic material removes itself and becomes active.

B The host cell continues to function and reproduce normally, making copies of the virus's genetic material as well as its own.

Viral Diseases

You might know that viruses cause many human diseases, such as chicken pox, influenza, some forms of pneumonia, and the common cold. But viruses also infect animals, causing diseases such as rabies and parvo. They can infect plants as well—in some cases causing millions of dollars of damage to crops. The tulips shown at the beginning of this lesson were infected with a virus that caused a streaked appearance on the petals. Most viruses attack and destroy specific cells. This destruction of cells causes the symptoms of the disease.

Some viruses cause symptoms soon after infection. Influenza viruses that cause the flu infect the cells lining your respiratory system, as shown in **Figure 14.** The viruses begin to replicate immediately. Flu symptoms, such as a runny nose and a scratchy throat, usually appear within 2–3 days.

Other viruses might not cause symptoms right away. These viruses are sometimes called latent viruses. Latent viruses continue replicating without damaging the host cell. HIV (human immunodeficiency virus) is one example of a latent virus that might not cause immediate symptoms.

HIV infects white blood cells, which are part of the immune system. Initially, infected cells can function normally, so an HIV-infected person might not appear sick. However, the virus can become active and destroy cells in the body's immune system, making it hard to fight other infections. It can often take a long time for symptoms to appear after infection. People infected with latent viruses might not know for many years that they have been infected.

Reading Check Why is HIV considered a latent virus?

The Flu 🔑

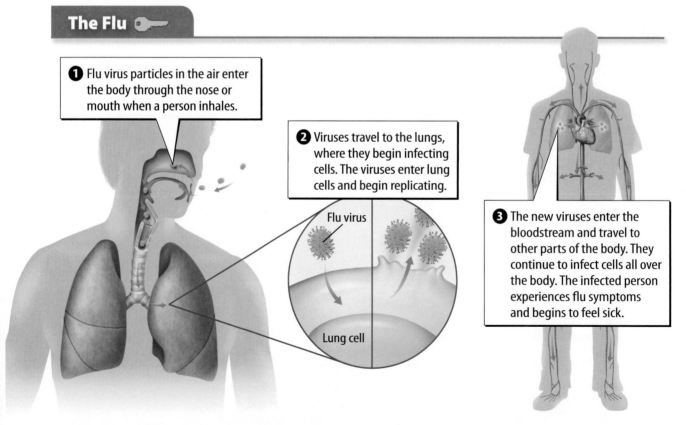

❶ Flu virus particles in the air enter the body through the nose or mouth when a person inhales.

❷ Viruses travel to the lungs, where they begin infecting cells. The viruses enter lung cells and begin replicating.

Flu virus

Lung cell

❸ The new viruses enter the bloodstream and travel to other parts of the body. They continue to infect cells all over the body. The infected person experiences flu symptoms and begins to feel sick.

Figure 14 Viruses that infect the respiratory system usually enter through the nose or mouth.

Visual Check Where do flu viruses replicate?

Treating and Preventing Viral Diseases

Since viruses are constantly changing, viral diseases can be difficult to treat. Antibiotics work only against bacteria, not viruses. Antiviral medicines can be used to treat certain viral diseases or prevent infection. These medicines prevent the virus from entering a cell or stop the virus from replicating. Antiviral medicines are specific to each virus. Like bacteria, viruses can rapidly change and become resistant to medicines.

Health officials use many methods to prevent the spread of viral diseases. One of the best ways to prevent a viral infection is to limit contact with an infected human or animal. The most important way to prevent infections is to practice good hygiene, such as washing your hands.

Immunity

Has anyone you know ever had chicken pox? Did they get it more than once? Most people who became infected with chicken pox develop an immunity to the disease. This is an example of acquired **immunity.** When a virus infects a person, his or her body begins to make special proteins called antibodies. An **antibody** *is a protein that can attach to a pathogen and make it useless.* Antibodies bind to viruses and other pathogens and prevent them from attaching to a host cell, as shown in **Figure 15.** The antibodies also target viruses and signal the body to destroy them. These antibodies can multiply quickly if the same pathogen enters the body again, making it easier for the body to fight infection. Another type of immunity, called natural immunity, develops when a mother passes antibodies on to her unborn baby.

WORD ORIGIN

immunity
from Latin *immunis*, means "exempt, free"

Antibodies

Figure 15 Antibodies bind to pathogens and prevent them from attaching to cells.

☑ **Visual Check** How does the antibody prevent the virus from attaching to the host cell?

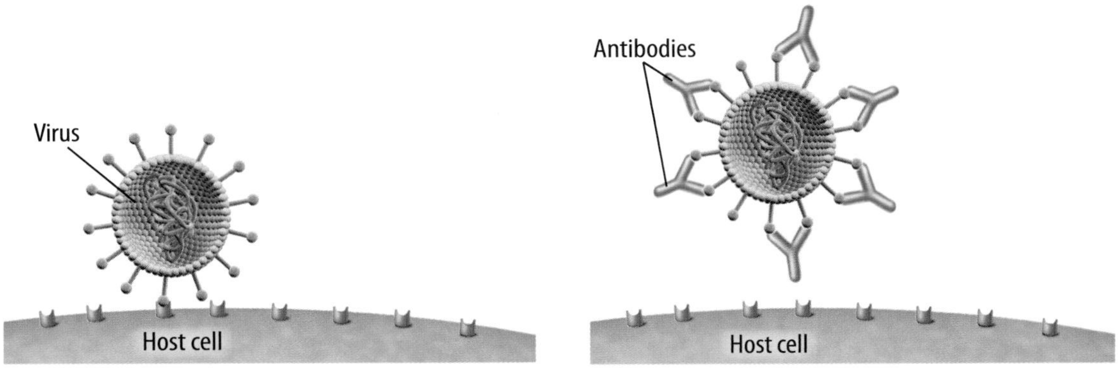

Antibodies

Virus

Host cell

Host cell

How do antibodies work?

When a virus infects a cell it binds to part of that cell called a receptor. The virus and the receptor fit together like puzzle pieces.

1. Read and complete a lab safety form.

2. Cut out two **virus shapes** and two **cell shapes.**

3. Using one virus shape and one cell shape, note how the virus fits against the receptor on the cell. **Tape** the virus and the cell together.

4. Cut out one **antibody shape.** Note how the virus shapes and the antibody shapes attach and tape them together.

5. Try to attach the virus shapes and the antibody shapes you just joined to the cell receptor.

Analyze and Conclude

1. **Observe** whether the virus or the joined virus and antibody were better able to attach to the cell.

2. **Key Concept** Explain how producing more antibodies would be beneficial during a viral infection.

Vaccines

One way to prevent viral diseases is through vaccination. *A vaccine is a mixture containing material from one or more deactivated pathogens, such as viruses.* When an organism is given a vaccine for a viral disease, the vaccine triggers the production of antibodies. This is similar to what would happen if the organism became infected with the virus normally. However, because the vaccine contains deactivated pathogens, the organism suffers only mild symptoms or none at all. After being vaccinated against a particular pathogen, the organism will not get as sick if exposed to the pathogen again.

Vaccines can prevent diseases in animals as well as humans. For example, pet owners and farmers get annual rabies vaccinations for their animals. This protects the animals from the disease. Humans are then protected from rabies.

Research with Viruses

Scientists are researching new ways to treat and prevent viral diseases in humans, animals, and plants. Scientists are also studying the link between viruses and cancer. Viruses can cause changes in a host's DNA or RNA, resulting in the formation of tumors or abnormal growth. Because viruses can change very quickly, scientists must always be working on new ways to treat and prevent viral diseases.

You might think that all viruses are harmful. However, scientists have also found beneficial uses for viruses. Viruses may be used to treat genetic disorders and cancer using gene transfer. Scientists use viruses to insert normal genetic information into a specific cell. Scientists hope that gene transfer will eventually be able to treat genetic disorders that are caused by one gene, such as cystic fibrosis or hemophilia.

Key Concept Check How do viruses affect human health?

Lesson 3 Review

Visual Summary

A virus is a strand of DNA or RNA surrounded by a layer of protein.

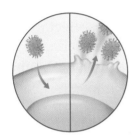

Viruses cause human diseases such as chicken pox and influenza.

A person's body produces proteins called antibodies that prevent an infection by viruses.

FOLDABLES®

Use your lesson Foldable to review the lesson. Save your Foldable for the project at the end of the chapter.

What do you think NOW?

You first read the statements below at the beginning of the chapter.

5. Viruses are the smallest living organisms.

6. Viruses can replicate only inside an organism.

Did you change your mind about whether you agree or disagree with the statements? Rewrite any false statements to make them true.

Use Vocabulary

1 **List** the different shapes a virus can have.

2 **Describe** in your own words how a vaccine works.

3 **Use the term** *antibodies* in a sentence.

Understand Key Concepts

4 **Describe** the structure of a virus.

5 Which is made by the body to fight viruses?
   A. antibody   **C.** bacteriophage
   B. bacteria   **D.** proteins

6 **Classify** a virus as a living or nonliving thing. Explain your answer.

7 **Compare** a vaccine and an antibody.

Interpret Graphics

8 **Draw** a graphic organizer like the one below including the steps that occur when a virus infects a cell.

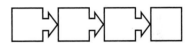

9 **Describe** what happens during this step of viral replication.

Critical Thinking

10 **Predict** the effect of preventing future mutations of the influenza virus.

11 **Evaluate** the importance of vaccines in keeping people healthy.

Bacterial Growth and Disinfectants

Materials

agar plates

cotton swabs

cellophane tape

permanent marker

hand sanitizer

Safety

Recall that pathogens such as bacteria and viruses are all around you. When studying pathogens, scientists often use agar plates to grow bacteria and other colonies. An agar plate is a Petri dish containing agar, a gel made from seaweed, and nutrients needed for bacteria to grow. When bacteria are transferred to an agar plate, they reproduce. After a few days, you can see colonies of bacteria. Disinfectants are chemicals that deactivate or kill pathogens such as bacteria. In this lab you will test how hand sanitizer, a common disinfectant, affects the growth of bacteria on agar plates.

Ask a Question

What effect does hand sanitizer have on bacterial growth?

Make Observations

1. Read and complete a lab safety form.

2. Set two agar plates on your desk or work area. Turn your agar plates upside down without opening them. With a permanent marker, label one plate *No Treatment* and the other *Disinfected*. Also write your name and the date on the plate. Turn the agar plates right side up.

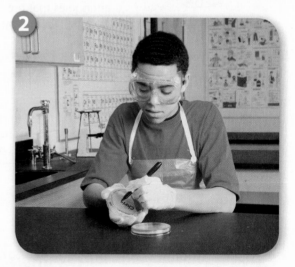

3. Rub the end of a cotton swab across the top of your desk or work area. Open the lid of the agar plate labeled *No Treatment* only enough to stick the swab in. Quickly make several S-shaped streaks on the agar. Close your plate and tape it shut.

4. Carefully clean the top of your desk or work area with hand sanitizer. Repeat step 3 using the agar plate labeled *Disinfected*.

5. Move your plates to an incubation area as directed by your teacher.

Form a Hypothesis

6 Using what you know about bacteria and disinfectants, write a hypothesis about how disinfectants affect the growth of bacteria. Make a prediction about how much bacterial growth you expect to see on your two agar plates.

Test Your Hypothesis

7 Check your agar plates after about three days. Record your observations in your Science Journal.

8 Compare the growth of bacteria on your two agar plates. Do your results support your hypothesis?

Analyze and Conclude

9 **Compare** Describe the differences in the amount of bacteria that grew on the two agar plates. Which plate had more?

10 What can you do to decrease the spread of bacteria in school and at home?

11 **Infer** Why didn't your experiment show any evidence of viral replication? How would you study the effect of disinfectants on viruses?

12 **The Big Idea** Why do doctors wash their hands or use hand sanitizer between appointments with different patients?

Communicate Your Results

Make a short video presentation about the results of your lab. Describe the question you investigated, the steps you took to answer your question, and the results that support your conclusions. Show your video to the class.

Inquiry Extension

Think about other situations in which cleanliness is important for preventing disease. Write a procedure in which you could test for bacteria as a comparison. Conduct your experiment and present your results to the class.

4

Lab Tips

☑ When streaking bacteria on your plates, use a steady, but light, pressure.

☑ After you disinfect your object, wait for the disinfectant to dry before testing the area.

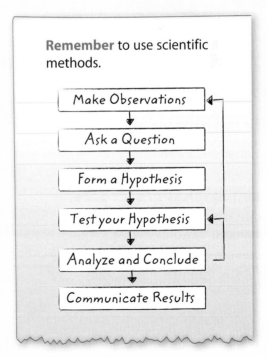

Remember to use scientific methods.

Make Observations
↓
Ask a Question
↓
Form a Hypothesis
↓
Test your Hypothesis
↓
Analyze and Conclude
↓
Communicate Results

Chapter 7 Study Guide

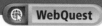

 Bacteria are unicellular prokaryotes, and viruses are small pieces of DNA or RNA surrounded by protein. Both bacteria and viruses can cause harmful diseases or can be useful to humans.

Key Concepts Summary

Vocabulary

Lesson 1: What are bacteria?

- **Bacteria** and archeans are unicellular organisms without nuclei. They have structures for movement, obtaining food, and reproduction.

- Bacteria exchange genetic information in a process called **conjugation.** They reproduce asexually by **fission.**

bacterium p. 231
flagella p. 234
fission p. 234
conjugation p. 234
endospore p. 235

Lesson 2: Bacteria in Nature

- Bacteria decompose materials, play a role in the nitrogen cycle, clean the environment, and are used in food.

- Some bacteria cause disease, while others are used to treat it.

decomposition p. 240
nitrogen fixation p. 240
bioremediation p. 241
pathogen p. 242
antibiotic p. 242
pasteurization p. 243

Lesson 3: What are viruses?

- A **virus** is made up of DNA or RNA surrounded by a protein coat.

- Viruses can cause disease, can be made into **vaccines,** and are used in research.

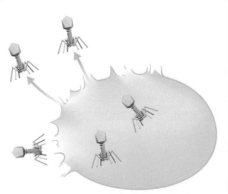

virus p. 247
antibody p. 251
vaccine p. 252

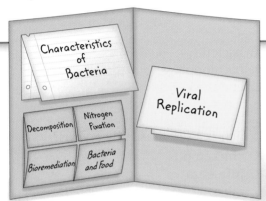
FOLDABLES® Chapter Project

Assemble your lesson Foldables as shown to make a Chapter Project. Use the project to review what you have learned in this chapter.

Characteristics of Bacteria

Viral Replication

Decomposition

Nitrogen Fixation

Bioremediation

Bacteria and Food

Use Vocabulary

1 Some bacteria have whiplike structures called _____ that are used for movement.

2 Your body produces proteins called _____ in response to infection by a virus.

3 Organisms that cause diseases are known as _____.

4 The process of killing bacteria in a food product by heating it is called _____.

5 Bacteria can form a(n) _____ to survive when environmental conditions are severe.

6 A(n) _____ is made by using pieces of deactivated viruses or dead pathogens.

Link Vocabulary and Key Concepts

((◎ Concepts in Motion) Interactive Concept Map

Copy this concept map, and then use vocabulary terms from the previous page and other terms from the chapter to complete the concept map.

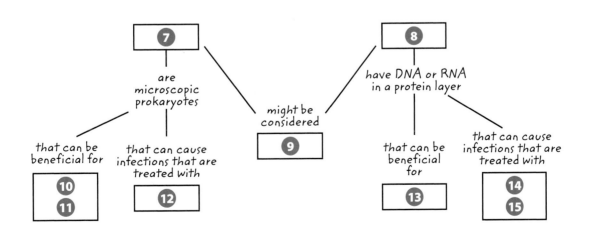

Understand Key Concepts

1 Which structure is NOT found in a bacterium?

A. chromosome
B. cytoplasm
C. nucleus
D. ribosome

2 Which structure helps a bacterium move?

A. capsule
B. endospore
C. flagellum
D. plasmid

3 What process is occurring in the illustration below?

A. budding
B. conjugation
C. fission
D. replication

4 Which term describes how bacteria can be used to clean up environmental waste?

A. bioremediation
B. decomposition
C. pasteurization
D. nitrogen fixation

5 Which statement correctly describes pathogens?

A. They are always bacteria.
B. They are in your body only when you are sick.
C. They break down dead organisms.
D. They cause disease.

6 Which statement correctly describes antibiotics?

A. They can kill any kind of bacterium.
B. They help bacteria grow.
C. They stop the growth and reproduction of bacteria.
D. They treat all diseases.

7 What is shown below?

A. bacteria
B. bacteriophage
C. endospore
D. virus

8 Which is NOT caused by a virus?

A. chicken pox
B. influenza
C. rabies
D. tuberculosis

9 What do vaccines stimulate the production of?

A. antibodies
B. DNA or RNA
C. protein
D. ribosomes

10 Scientists hope to be able to use viruses for gene therapy because viruses can

A. become latent for long periods of time.
B. inject genetic material into host cells.
C. make proteins to attack cells.
D. transport themselves throughout the body.

11 Which statement correctly describes viruses?

A. All viruses are latent.
B. All viruses contain DNA.
C. Viruses are considered living things.
D. Viruses do not have organelles.

Critical Thinking

12 **Compare and contrast** bacteria and archaea.

13 **Evaluate** the importance of bacterial conjugation.

14 **Model** the life of a bacterium that performs nitrogen fixation in the soil.

15 **Contrast** asexual reproduction in bacteria and replication in viruses. What are some advantages and disadvantages of each?

16 **Organize** the effects of bacteria on health by copying and completing the table below.

Harmful Effects	Beneficial Effects

17 **Analyze** the importance of vaccines in preventing large outbreaks of influenza.

18 **Draw** and label a typical bacterium. Are the features you labeled beneficial for moving, for finding food, or for another purpose? Explain your answer.

19 **Explain** what happens during the process shown below. How does this process eventually create new strains of bacteria that are resistant to antibiotics?

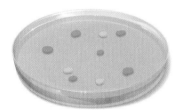

Writing in Science

20 **Summarize** an argument that you could use to encourage all the families in your neighborhood to make sure their pets are vaccinated against rabies.

REVIEW THE BIG IDEA

21 What are bacteria and viruses and why are they important? Include examples of how they are both beneficial and harmful to humans.

22 Describe what is happening in the photo below. Explain what is happening to both the bacterium and the virus.

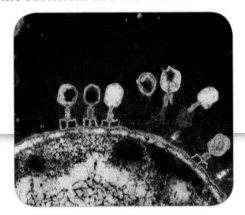

Math Skills ×÷

Review

— Math Practice —

Use a Formula

23 How many bacteria would there be if 100 bacteria underwent fission 8 times?

24 If each fission cycle takes 20 minutes, how many cycles would it take for 100 bacteria to divide into 100,000?

25 A strain of bacteria takes 30 minutes to undergo fission. Starting with 500 bacteria, how many would there be after 4 hours?

Standardized Test Practice

Record your answers on the answer sheet provided by your teacher or on a sheet of paper.

Multiple Choice

1 Which is NOT a characteristic of bacteria?

  A They are microscopic.

  B They are unicellular.

  C They can live in many environments.

  D They have a membrane-bound nucleus.

2 Which process increases genetic diversity in bacteria?

  A attachment to a host organism

  B division into two organisms

  C formation of an endospore

  D transfer of plasmid strands

Use the diagram below to answer questions 3 and 4.

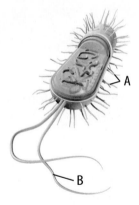

3 The diagram above illustrates a bacterium. What is the function of the structure labeled *A?*

  A attaching to surfaces

  B sensing surroundings

  C stinging prey

  D taking in nutrients

4 The structure labeled *B* helps a bacterium

  A move.

  B protect itself.

  C reproduce.

  D transfer DNA.

5 What beneficial vitamin do some human intestinal bacteria produce?

  A vitamin A

  B vitamin C

  C vitamin D

  D vitamin K

6 Which statement BEST explains why living organisms in an ecosystem depend on bacteria?

  A Bacteria help reduce the number of predators.

  B Bacteria kill weaker members of a species so only the stronger ones survive.

  C Bacteria protect organisms from harmful solar rays.

  D Bacteria release molecules into soil that are used by other organisms.

Use the diagram below to answer question 7.

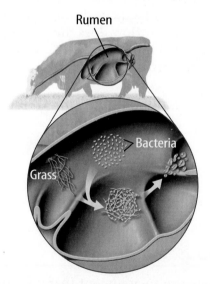

7 What role do bacteria play in the process shown above?

  A They break down cellulose.

  B They convert nitrogen in grass.

  C They prevent viruses from growing.

  D They remove harmful pollutants.

✓ Assessment

Online Standardized Test Practice

8 In which process do bacteria and other organisms clean up environmental pollution?

 A bioremediation

 B decomposition

 C fixation

 D pasteurization

Use the diagram below to answer question 9.

9 What is pictured in the diagram above?

 A an antibody

 B a bacteriophage

 C a bacterium

 D a plasmid

10 Which BEST explains how mutation benefits a virus?

 A It enables the virus to adjust to changes in its host cell.

 B It enables the virus to reproduce more quickly.

 C It enables the virus to resist antibiotic therapy.

 D It enables the virus to travel from host to host.

Constructed Response

Use the diagrams below to answer questions 11 and 12.

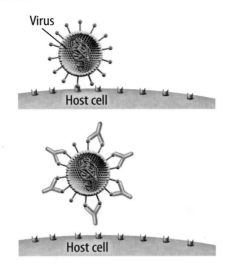

11 Describe how the virus attaches to the host cell in the figure at the top of the diagram.

12 What are the Y-shaped structures on the virus in the figure at the bottom of the diagram? Explain their interaction with the virus.

13 Why can viral infections be more difficult to treat than bacterial infections?

14 What are two methods you can use to prevent a viral infection?

15 What happens to the host cell when a latent virus goes through an inactive stage?

NEED EXTRA HELP?															
If You Missed Question...	1	2	3	4	5	6	7	8	9	10	11	12	13	14	15
Go to Lesson...	1	1	1	1	2	2	2	2	3	3	3	3	3	3	3

Protists and Fungi

THE BIG IDEA

What are protists and fungi, and how do they affect an environment?

Plant or Animal?

These organisms are neither! Protists and fungi are two groups of living things that have characteristics similar to those of plants or animals.

- How is the organism pictured similar to a plant? An animal?

- How might this organism benefit its environment?

Get Ready to Read

What do you think?

Before you read, decide if you agree or disagree with each of these statements. As you read this chapter, see if you change your mind about any of the statements.

1 Protists are grouped together because they all look similar.

2 Some protists cause harm to other organisms.

3 Many protists make their own food.

4 Mushrooms and yeasts are two types of fungi.

5 Fungi are always helpful to plants.

6 Some fungi can be made into foods or medicines.

ConnectED Your one-stop online resource

connectED.mcgraw-hill.com

🎬 Video | 🌐 WebQuest

🔊 Audio | ✓ Assessment

📑 Review | 🔊 Concepts in Motion

❓ Inquiry | g Multilingual eGlossary

Lesson 1

Reading Guide

Key Concepts 🔑
ESSENTIAL QUESTIONS

- What are the different types of protists and how do they compare?
- How are protists beneficial?

Vocabulary
protist p. 265
algae p. 266
diatom p. 267
protozoan p. 270
cilia p. 270
paramecium p. 270
amoeba p. 271
pseudopod p. 271

g   **Multilingual eGlossary**

▢ **Video** **BrainPOP®**

What are protists?

 Grabbing a Snack?

The protist group includes diverse organisms. What is the blue-tinted organism doing in the photo? How is this organism similar to an animal?

How does a protist react to its environment?

Like other organisms, protists can react to their environment in many ways. One type of protist called *Euglena* has specialized structures to move, perform photosynthesis, and react to light.

1 Read and complete a lab safety form.

2 Place a **Petri dish** containing a ***Euglena* culture** on a white piece of **paper.** Using a **hand lens,** observe the *Euglena.*

3 Carefully cut a hole the size of a dime in a piece of **aluminum foil.** Place the foil on top of the dish so that the hole is centered over the top. Shine the light from a **desk lamp** at the hole.

4 At the end of class, remove the foil and observe the *Euglena* again.

Think About This

1. Where were the *Euglena* in the dish at the beginning of class? At the end?

2. Why do you think this behavior is beneficial to *Euglena*?

3. 🔑 **Key Concept** What structures do you think help *Euglena* react to its environment?

What are protists?

When you see a living thing, one of the first questions you might have is whether it is a plant or an animal. You might recognize a dog as an animal because of its fur. You might know a flower is a plant because of its leaves. Besides appearance, organisms can also be classified by structures in their cells. For example, a plant cell has a cell wall made of cellulose and a membrane made of flexible fats. A plant cell often contains chloroplasts, organelles that carry out photosynthesis. An animal cell also has a membrane made of flexible fats but does not contain chloroplasts or have a cell wall. These characteristics make it easy to identify both types of cells. However, some organisms, such as the protist shown in **Figure 1,** cannot be classified as easily.

A **protist** is *a member of a group of eukaryotic organisms, which have a membrane-bound nucleus.* Members of the protist group share some characteristics with plants, animals, or organisms known as fungi. However, they are not classified as any of these groups. Although protists are classified together, they are diverse and have different adaptations for movement and for finding food.

✓ **Reading Check** What is a protist?

Figure 1 Many photosynthetic algae look like plants.

Reproduction of Protists

Most protists reproduce asexually. What does the offspring of **asexual reproduction** look like? It is an exact copy of the parent. Asexual reproduction can create new organisms quickly. However, many protists can also reproduce sexually. Offspring of sexual reproduction are genetically different from the parents. Sexual reproduction takes more time, but it creates new organisms with a variety of characteristics.

Classification of Protists

Scientists usually classify organisms according to their similarities. However, protists are a unique and diverse classification of organisms. Typically, a protist is any eukaryote that cannot be classified as a plant, an animal, or a fungus. However, protists might look and act very much like these other types of organisms. Scientists classify protists as plantlike, animal-like, or funguslike based on which group they most resemble, as shown in **Table 1.**

Key Concept Check What are the different types of protists?

Review Personal Tutor

Table 1 Protists Classified into One of Three Groups

Classification	Plantlike	Animal-like	Funguslike
Example	algae 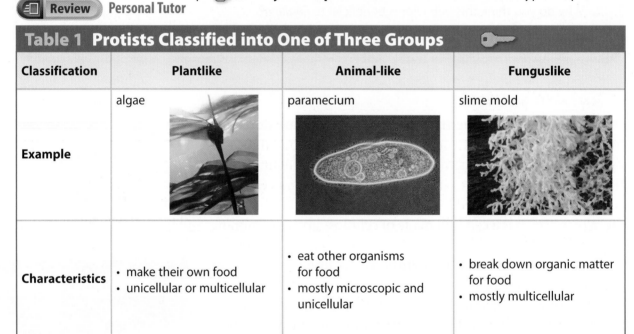	paramecium	slime mold
Characteristics	• make their own food • unicellular or multicellular	• eat other organisms for food • mostly microscopic and unicellular	• break down organic matter for food • mostly multicellular

Plantlike Protists

You might have seen brown, green, or red seaweed at the beach or in an aquarium. These seaweeds are algae (AL jee; singular, alga), one type of plantlike protist. Why might they be classified as plantlike? **Algae** *are plantlike protists that produce food through photosynthesis using light energy and carbon dioxide.* Most plantlike protists, however, are much smaller than the multicellular algae shown in **Table 1.** You can't see most algae without a microscope.

Diatoms

A type of microscopic plantlike protist with a hard outer wall is a **diatom** (DI uh tahm). Diatoms are so common that if you filled a cup with water from the surface of any lake or pond, you would probably collect thousands of them. Look at the unicellular diatoms shown at the top of **Figure 2.** A diatom can resemble colored glass. In fact, the cell walls of diatoms contain a large amount of silica, the main mineral in glass.

Dinoflagellates

Can you guess how the protist in the middle of **Figure 2** moves? This organism is a dinoflagellate (di noh FLA juh lat), a unicellular plantlike protist that has flagella—whiplike parts that enable the protist to move. The flagella beat back and forth, enabling the dinoflagellate to spin and turn. Some of these protists glow in the dark because of a chemical reaction that occurs when they are disturbed.

 Reading Check What purpose do flagella serve?

Euglenoids

Another type of plantlike protist also uses flagella to move but has a unique structure covering its body. A euglenoid (yew GLEE noyd), shown at the bottom of **Figure 2,** is a unicellular plantlike protist with a flagellum at one end of its body. Instead of a cell wall, euglenoids have a rigid, rubbery cell coat called a pellicle (PEL ih kul). Euglenoids have eyespots that detect light and determine where to move. Euglenoids swim quickly and can creep along the surface of water when it is too shallow to swim. These protists have chloroplasts and make their own food. If there is not enough light for making food, they can absorb nutrients from decaying matter in the water. Animals such as tadpoles and small fish eat euglenoids.

 Reading Check What characteristics do plantlike protists share with plants?

Diatoms

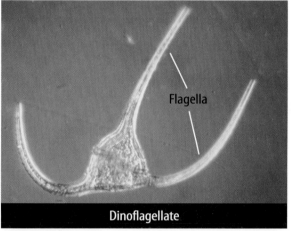
Flagella

Dinoflagellate

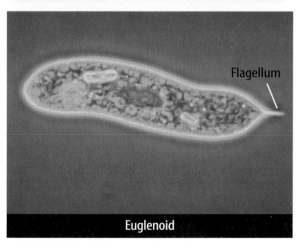
Flagellum

Euglenoid

Figure 2 All of these microscopic organisms are protists. The cell walls of diatoms contain silica. The dinoflagellate has two flagella that cause it to spin. The euglenoid has a flagellum and a rigid cell coat.

Algae

Recall that algae are photosynthetic plantlike protists. Some algae are big and multicellular, like the seaweeds in **Figure 3.** Other algae are unicellular and can be seen only with a microscope. Algae are classified as red, green, or brown, depending on the pigments they contain.

Some types of red and brown algae appear similar to plants. Unlike plants, these algae do not have a complex organ system for transporting water and nutrients. Instead of roots, they have holdfasts, structures that secrete a chemical-like glue that fastens them to the rocks.

One unusual green alga is volvox. In **Figure 3** you can see that many volvox cells come together to form a larger sphere. These cells move together as one group and beat their flagella in unison. Some cells produce parts necessary for sexual reproduction. The volvox cells in the front of the group have larger eyespots that sense light for photosynthesis. Do you think volvox should be considered unicellular or multicellular?

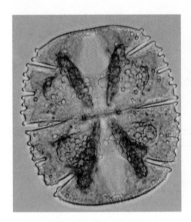

Unicellular Algae

Multicellular Algae

Figure 3 Volvox are unicellular green algae that join together to form a sphere. ▼

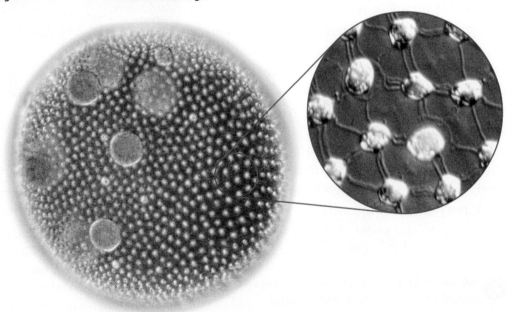
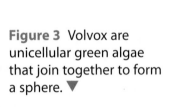

The Importance of Algae

Do you use algae in your everyday life? You might be surprised by all the materials you use that contain algae. You might be eating algae if you snack on ice cream, marshmallows, or pudding. Algae are a common ingredient in other everyday products, including toothpaste, lotions, fertilizers, and some swimming pool filters.

Algae and Ecosystems

Algae provide food for animals and animal-like protists. They also provide shelter for many aquatic organisms. In **Figure 4,** you can see that some brown algae grow tall. Thick groups of tall brown algae are called kelp forests. Sea otters and seals come to the kelp forest to eat smaller animals.

 Key Concept Check How are algae beneficial to an ecosystem?

Do you think algae ever cause problems in an ecosystem? Algae and other photosynthetic protists can help remove pollution from the water. However, this pollution can be a food source for the algae, allowing the population of algae to increase quickly. The algae produce wastes that can poison other organisms. As shown in **Figure 5,** when the number of these protists increase, the water can appear red or brown. This is called a red tide or a harmful algal bloom (HAB).

 Reading Check What causes a red tide?

▲ **Figure 4** Brown algae can form thick kelp forests that are home to many animals and other protists.

Figure 5 Red tides can be harmful to aquatic organisms. ▼

Animal-like Protists

Some protists are similar to plants, but others are more like animals. **Protozoans** (proh tuh ZOH unz) *are protists that resemble tiny animals.* Animal-like protists all share several characteristics. They do not have chloroplasts or make their own food. Protozoans are usually microscopic and all are unicellular. Most protozoans live in wet environments.

Ciliates

Cilia (SIH lee uh) *are short, hairlike structures that grow on the surface of some protists.* Protists that have these organelles are called ciliates. Cilia cover the surface of the cell. They can beat together and move the animal-like protist through the water.

✓ **Reading Check** What function do cilia perform?

A common protozoan with these cilia is the **paramecium** (pa ruh MEE see um; plural, paramecia)—*a protist with cilia and two types of nuclei.* One example of a paramecium is shown in **Figure 6.** A paramecium, like most ciliates, gets its food by forcing water into a groove in its side. The groove closes and a food vacuole, or storage area, forms within the cell. The food particles are digested and the extra water is forced back out. Ciliates reproduce asexually, but they can exchange some genetic material through a **process** called conjugation (kahn juh GAY shun). This results in more genetic variation.

ACADEMIC VOCABULARY

process
(noun) an event marked by gradual changes that lead toward a particular result

Paramecium

((O **Concepts in Motion** Animation

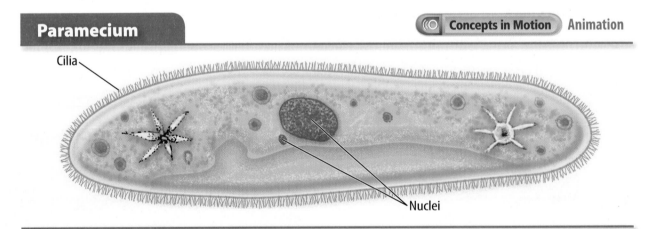

Cilia

Nuclei

Figure 6 A paramecium, like the one shown above, has two nuclei and is covered with hairlike structures called cilia.

Flagellates

Recall that dinoflagellates, a type of plant-like protist, use one or more flagella to move. A type of protozoan also has one or more flagella—a flagellate. However, a flagellate does not always spin when it moves.

Flagellates eat decaying matter including plants, animals, and other protists. Many flagellates live in the digestive system of animals and absorb nutrients from food eaten by them.

Sarcodines

Animal-like protists called sarcodines (SAR kuh dinez) have no specific shape. At rest, a sarcodine resembles a random cluster of cytoplasm, or cellular material. These animal-like protists can ooze into almost any shape as they slide over mud or rocks.

An **amoeba** (uh MEE buh) *is one common sarcodine with an unusual adaptation for movement and getting nutrients.* An amoeba moves by using a **pseudopod,** *a temporary "foot" that forms as the organism pushes part of its body outward.* It moves by first stretching out a pseudopod then oozing the rest of its body up into the pseudopod. This movement is shown in **Figure 7.**

Amoebas also use pseudopods to get nutrients. An amoeba surrounds a smaller organism or food particle with its pseudopod and then oozes around it. A food vacuole forms inside the pseudopod where the food is quickly digested. You can see an amoeba capturing its prey in the photo at the beginning of this lesson.

Some sarcodines get nutrients and energy from ingesting other organisms, while others make their own food. Some sarcodines even live in the digestive systems of humans and get nutrients and energy from the human's body.

Amoeba Movement

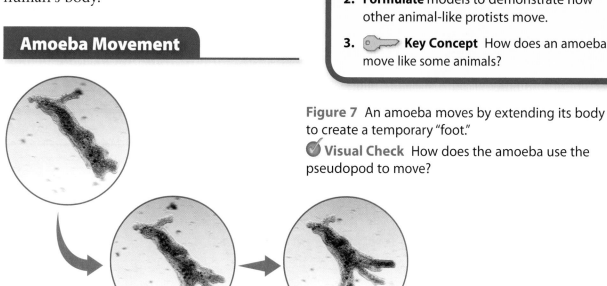

Inquiry MiniLab
15 minutes

How can you model the movement of an amoeba?

The way an amoeba moves is so unusual that scientists use the term to describe a specific type of movement. Organisms that move by oozing are said to have "amoeboid" movement.

1. Read and complete a lab safety form.
2. Half-fill a **sock** with **dry beans.** Tie the end of the sock closed with a piece of **string.**
3. Place the sock on a flat surface and spread the beans evenly within the sock.
4. Demonstrate the organism's movement by pushing the beans forward in the sock until the sock moves.

5. Model how amoebas capture their food by pushing the beans and sock around your finger.

Analyze and Conclude

1. **Explain** how amoebas capture their food.
2. **Formulate** models to demonstrate how other animal-like protists move.
3. 🔑 **Key Concept** How does an amoeba move like some animals?

Figure 7 An amoeba moves by extending its body to create a temporary "foot."

✓ **Visual Check** How does the amoeba use the pseudopod to move?

Fold a sheet of paper to make a three-tab book. Label your book as shown. Use it to organize your notes about protozoans and how they move.

How Protozoans Move

| Flagella | Cilia | Pseudopod |

The Importance of Protozoans

Imagine living in a world without organisms that decompose other organisms. Plant material and dead animals would build up until the surface of Earth quickly became covered. Many protozoans are beneficial to an environment because they break down dead plant and animal matter. This decomposed matter is then recycled back into the environment and used by living organisms.

Some protozoans can cause disease by acting as parasites. These organisms can live inside a host organism and feed off it. Protozoan parasites are responsible for millions of human deaths every year.

One example of a disease caused by a protist is malaria. **Figure 8** illustrates how malaria develops and is spread to humans by mosquitoes. Protozoan parasites called plasmodia (singular, plasmodium) live and reproduce in red blood cells. Malaria kills more than one million people each year.

Key Concept Check In what ways are protists helpful and harmful to humans?

Plasmodium Life Cycle

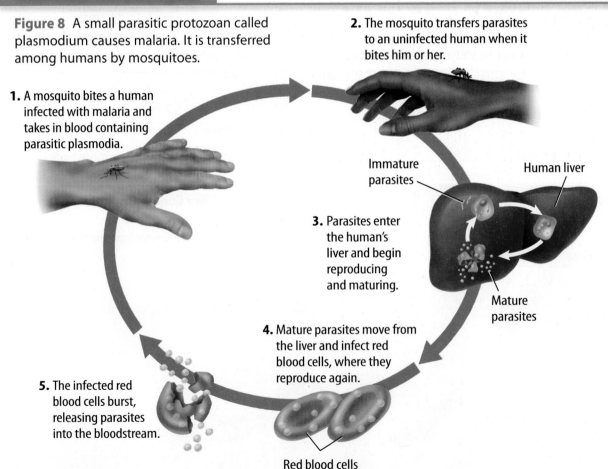

Figure 8 A small parasitic protozoan called plasmodium causes malaria. It is transferred among humans by mosquitoes.

1. A mosquito bites a human infected with malaria and takes in blood containing parasitic plasmodia.

2. The mosquito transfers parasites to an uninfected human when it bites him or her.

Immature parasites

Human liver

3. Parasites enter the human's liver and begin reproducing and maturing.

Mature parasites

4. Mature parasites move from the liver and infect red blood cells, where they reproduce again.

5. The infected red blood cells burst, releasing parasites into the bloodstream.

Red blood cells

Funguslike Protists

In addition to plantlike and animal-like protists, there are funguslike protists. These protists share many characteristics with fungi. However, because of their differences from fungi, they are classified as protists.

Slime and Water Molds

Have you ever seen a strange organism like the one shown in **Figure 9?** These funguslike protists, called slime molds, look like they could have come from another planet. The body of the slime mold is composed of cell material and nuclei floating in a slimy mass. Most slime molds absorb nutrients from other organic matter in their environment.

A Funguslike Protist

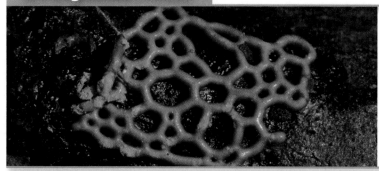

Figure 9 Slime molds come in a variety of colors and forms. These protists often live on the surfaces of plants.

A water mold is another kind of funguslike protist that lives as a parasite or feeds on dead organisms. Originally classified as fungi, water molds often cause diseases in plants.

Both slime molds and water molds reproduce sexually and asexually. The molds usually reproduce sexually when environmental conditions are harsh or unfavorable.

Importance of Funguslike Protists

Funguslike protists play a valuable role in the ecosystem. They break down dead plant and animal matter, making the nutrients from these dead organisms available for living organisms. While some slime molds and water molds are beneficial, many others can be very harmful.

Many funguslike protists attack and consume living plants. The Great Irish Potato Famine resulted from damage by a funguslike protist. In 1845 this water mold destroyed more than half of Ireland's potato crop. More than one million people starved as a result.

 Key Concept Check How are funguslike protists beneficial to an environment?

Lesson 1 Review

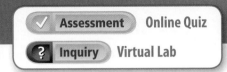
Visual Summary

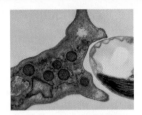

Protists are a diverse group of organisms that cannot be classified as plants, animals, or fungi.

Protists are grouped according to the type of organisms they most resemble. Diatoms are one type of plantlike protist.

Some protists use hairlike structures called cilia to move.

FOLDABLES

Use your lesson Foldable to review the lesson. Save your Foldable for the project at the end of the chapter.

What do you think NOW?

You first read the statements below at the beginning of the chapter.

1. Protists are grouped together because they all look similar.

2. Some protists cause harm to other organisms.

3. Many protists make their own food.

Did you change your mind about whether you agree or disagree with the statements? Rewrite any false statements to make them true.

Use Vocabulary

1 **Distinguish** between cilia and flagella.

2 **Define** *pseudopod* in your own words or with a drawing.

Understand Key Concepts 🔑

3 **List** three groups of animal-like protists and three groups of plantlike protists.

4 **Describe** one example of how protists benefit humans.

5 Identify which protist causes red tides.
A. algae   C. euglenoids
B. diatoms   D. paramecia

Interpret Graphics

6 **Identify** Copy and fill in the graphic organizer with the three categories of protists.

7 The image below shows the plasmodium life cycle.

Explain in your own words how this disease can be spread among people.

Critical Thinking

8 **Formulate** a plan for deciding to classify a newly discovered protist.

The Benefits of Algae

Big Benefits from Tiny Organisms

▲ Processing plants, such as this one, are a major source of algae oil.

Algae are protists that can do more than just cover a pond as slimy scum. They release oxygen through photosynthesis. In fact, most of the oxygen in Earth's atmosphere comes from photosynthesis that occurs in algae, plants, and some bacteria. Algae are also food for many organisms, including humans. But algae can provide something else very valuable—oil.

A microalga is another type of protist that is very small and reproduces quickly. The total mass of some microalgae can double several times a day. More than half of their mass is fats, also called lipids, that store energy. One type of lipid, triglycerides, can be turned into diesel oil, gasoline, and jet fuel.

Microalgae can grow outdoors in ponds and produce 100 times more oil per acre than any other crop. They also can grow indoors under lights in photobioreactors. A photobioreactor is a tank filled with water and nutrients. Photosynthesis requires carbon dioxide. Instead of releasing carbon dioxide gas into the atmosphere, power plants can pump it into photobioreactors for microalgae to use. Also, microalgae can grow in water that is unsafe to drink. Using this technology, microalgae can grow in areas, such as deserts, where it is not ordinarily possible to grow other crops.

It's Your Turn

RESEARCH Protists, including algae, are important sources of food. Research five types of organisms that depend on protists for food. Make a display of your results to share with your class.

What are fungi?

Reading Guide

Key Concepts 🔑
ESSENTIAL QUESTIONS

- What are the different types of fungi and how do they compare?

- Why are fungi important?

- What are lichens?

Vocabulary

hyphae p. 277

mycelium p. 277

basidium p. 278

ascus p. 279

zygosporangia p. 279

mycorrhizae p. 282

lichen p. 284

g Multilingual eGlossary

The organism pictured is a puffball mushroom, named for the puff of material that it releases. What do you think the material is? What is the purpose of the puff of material?

Is there a fungus among us?

The mold you see on food is fungi that are consuming and decomposing it. Fungi are also found as molds or mushrooms on wood, mulch, and other organic materials.

1. Read and complete a lab safety form.
2. Examine the different **samples of fungi** your teacher provides. Use a **magnifying lens** to observe similarities and differences among the samples.
3. Record your observations in your Science Journal. Include drawings of the different structures or characteristics you notice.

Think About This

1. What similarities did you see among the fungi samples?

2. Why do you think your teacher had the mold samples in closed containers?

3. **Key Concept** In what ways do you think the fungi you observed are helpful or not helpful to people?

What are fungi?

What would you guess is the world's largest organism? A fungus in Oregon is the largest organism ever measured by scientists. It stretches almost 9 km². Fungi, like protists, are eukaryotes. Scientists estimate more than 1.5 million species of fungi exist.

Fungi form long, threadlike structures that grow into large tangles, usually underground. *These structures, which absorb minerals and water, are called* **hyphae** (HI fee). *The hyphae create a network called the* **mycelium** (mi SEE lee um), shown in **Figure 10.** The fruiting body of the mushroom, the part above ground, is also made of hyphae.

Fungi are heterotrophs, meaning they cannot make their own food. Some fungi are parasites, obtaining nutrients from living organisms. Fungi dissolve their food by releasing chemicals that decompose organic matter. Fungi then absorb the nutrients.

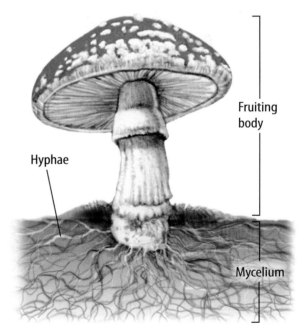

Figure 10 Mushrooms are common fungi. In the drawing, you can see mycelium, the network of hyphae. The hyphae release enzymes and absorb water and nutrients.

Hyphae

Fruiting body

Mycelium

Types of Fungi

Scientists group fungi based on how they look and how they reproduce. Although fungi can reproduce sexually or asexually, almost all reproduce asexually by producing spores. Spores are small reproductive cells with a strong, protective outer covering. The spores can grow into new individuals.

The classification of fungi often changes as scientists learn more about them. Today, scientists recognize four groups of fungi: club fungi, sac fungi, zygote fungi, and imperfect fungi. As technology helps scientists understand more about fungi, the categories might change.

 Key Concept Check What are the four groups of fungi?

Club Fungi

When you think of fungi, you might think of a **mushroom**. Mushrooms belong to the group called club fungi. They are named for the clublike shape of their reproductive structures. However, the mushroom is just one part of the fungus. The part of the mushroom that grows above ground is a structure called a basidiocarp (bus SIH dee oh karp). Inside the basidiocarp are the **basidia** (buh SIH dee uh; singular, basidium), *reproductive structures that produce sexual spores.* Most of a club fungus is a network of hyphae that grows underground and absorbs nutrients.

✓ **Reading Check** Where does most of a club fungus grow?

Many club fungi are named for their various shapes and characteristics. Club fungi include puffballs like those at the beginning of the lesson, stinkhorns, and the bird's nest fungi shown in **Figure 11.** There is even a club fungus that glows in the dark due to a chemical reaction in its basidiocarp.

Figure 11 Club fungi, such as this bird's nest fungus, use basidiospores to reproduce.

✓ **Visual Check** Which part of the fungus is club-shaped?

Club Fungi 🔑

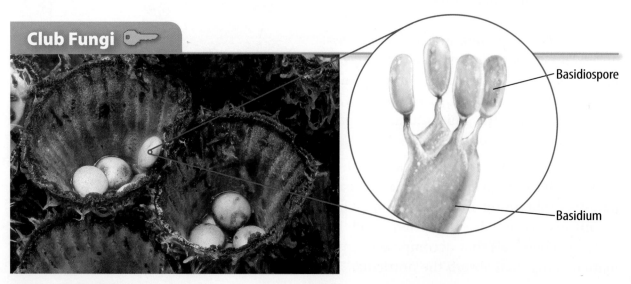

Basidiospore

Basidium

Sac Fungi

Do you know what bread and a diaper rash have in common? A type of sac fungus causes bread dough to rise. A different sac fungus is responsible for a rash that babies can develop on damp skin under their diapers. Many sac fungi cause diseases in plants and animals. Other common sac fungi, such as truffles and morels, are harvested by people for food.

Like club fungi, sac fungi are named for their reproductive structures. *The* **ascus** (AS kuhs; plural, asci) *is the reproductive structure where spores develop on sac fungi.* The ascus often looks like the bottom of a tiny bag or sack. The spores from sac fungi are called ascospores (AS kuh sporz). Sac fungi can undergo both sexual and asexual reproduction. Many yeasts are sac fungi, including the common yeast used to make bread, as shown in **Figure 12.** When the yeast is mixed with water and warmed, the yeast cells become active. They begin cellular respiration and release carbon dioxide gas. This causes the bread dough to rise.

Zygote Fungi

Another type of fungus can cause bread to develop mold. Bread mold, like the type shown in **Figure 12,** is caused by a type of fungus called a zygote fungus. You might also find zygote fungi growing in moist areas, such as a damp basement or on a bathroom shower curtain.

The hyphae of a zygote fungus grow over materials, such as bread, dissolving the material and absorbing nutrients. *Tiny stalks called* **zygosporangia** (zi guh spor AN jee uh) *form when the fungus undergoes sexual reproduction.* The zygosporangia release spores called zygospores. These zygospores then fall on other materials where new zygote fungi might grow.

Reading Check How do sac and zygote fungi differ?

Figure 12 Some fungi can be used to make food, but other fungi can eat the food too.

Zygosporangia

Fold a sheet of paper to make a four-door book. Label it as shown. Use your book to organize information about the characteristics of the different classifications of fungi.

Zygote fungi | Sac fungi

Club fungi | Imperfect fungi

Imperfect Fungi

How are itchy feet and blue cheese connected? They can both be caused by imperfect fungi. You might have had athlete's foot, an infection that causes flaking and itching in the skin of the feet. The imperfect fungus that causes athlete's foot grows and reproduces easily in the moist environment near a shower or in a sweaty shoe. The blue color you see in blue cheese comes from colonies of a different type of imperfect fungi. They are added to the milk or the curds during the cheese-making process.

Imperfect fungi are named because scientists have not observed a sexual, or "perfect," reproductive stage in their life cycle. Since fungi are classified according to the shape of their reproductive structures, these fungi are left out, or labeled "imperfect." Often after a species of imperfect fungi is studied, the sexual stage is observed. The fungi is then classified as a club, sac, or zygote fungus based on these observations.

✔ **Reading Check** Why are imperfect fungi classified that way?

Inquiry MiniLab

10 minutes

What do fungal spores look like?

Have you ever seen spores from a mushroom? Some types of fungi reproduce by releasing these structures.

1. Read and complete a lab safety form.
2. Carefully remove the stem of your **mushroom.** Observe the gills, the soft structures on the underside of the cap.
3. Gently place the mushroom cap with the gills down on a sheet of **unlined white paper.**
4. Let the mushroom cap sit undisturbed overnight. Remove it from the paper the next day.

Analyze and Conclude

1. **Describe** your results and sketch them in your Science Journal. What caused this result?

2. **Estimate** the number of mushrooms that could be produced from a single mushroom cap.

3. **Key Concept** What type of fungi (club, sac, zygote, or imperfect) did you use to make the print?

Figure 13 Products such as bread, cheese, and medicines are made using fungi.

The Importance of Fungi

Do you like chocolate, carbonated sodas, cheese, or bread? If so, you might agree that fungi are beneficial to humans. Fungi are involved in the production of many foods and other products, as shown in **Figure 13.** Some fungi are used as a meat substitute because they are high in protein and low in cholesterol. Other fungi are used to make antibiotics.

Decomposers

Fungi help create food for people to eat, but they are also important because of the things they eat. As you read earlier, fungi are an important part of the environment because they break down dead plant and animal matter, as shown in **Figure 14.** Without fungi and other decomposers, dead plants and animals would pile up year after year. Fungi also help break down pollution, including pesticides, in soil. Without fungi to destroy it, pollution would build up in the environment.

Living things need nutrients. The nutrients available in the soil would eventually be used up if they were not replaced by decomposing plant and animal matter. Fungi help put these nutrients back into the soil for plants to use.

Figure 14 Fungi help decompose dead organic matter. ▼

Math Skills

Using Fractions

Under certain conditions, 100 percent of the cells in fungus A reproduce in 24 hours. The number of cells of fungus A doubles once each day.

Day 1 = 10,000 cells

Day 2 = 20,000 cells

Day 3 = 40,000 cells

Day 4 = 80,000 cells

When an antibiotic is added to the fungus, the growth is reduced by 50 percent. Only half the cells reproduce each day.

Day 2 = 15,000 cells

Day 3 = 22,500 cells

Day 4 = 33,750 cells

Practice

Without an antibiotic, how many cells of fungus A would there be on day 6?

 Review

- **Math Practice**
- **Personal Tutor**

Fungi and Plant Roots

Plants benefit from fungi in other ways, too. Many fungi and plants grow together, helping each other. Recall that fungi take in minerals and water through the hyphae, or threadlike structures that grow on or under the surface. *The roots of the plants and the hyphae of the fungi weave together to form a structure called* **mycorrhiza** (mi kuh RI zuh; plural, micorrhizae).

Mycorrhizae can exchange molecules, as shown in **Figure 15.** As fungi break down decaying matter in the soil, they make nutrients available to the plant. They also increase water absorption by increasing the surface area of the plant's roots.

Fungi cannot photosynthesize, or make their own food using light energy. Instead, the fungi in mycorrhizae take in some of the sugars from the plant's photosynthesis. The plants benefit by receiving more nutrients and water. The fungi benefit and continue to grow by using plant sugars. Scientists suspect that most plants gain some benefit from mycorrhizae.

✓ **Reading Check** How do mycorrhizae benefit both the plant and the fungus?

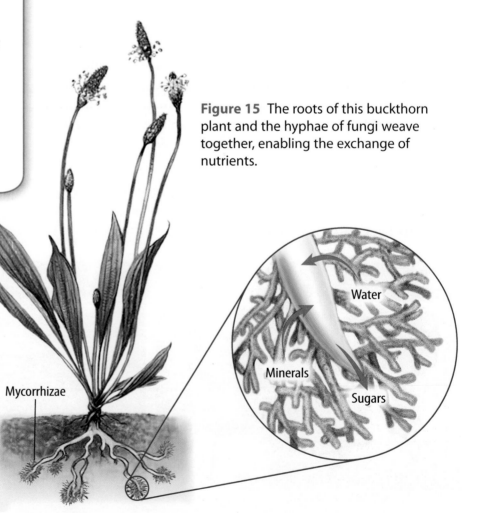

Figure 15 The roots of this buckthorn plant and the hyphae of fungi weave together, enabling the exchange of nutrients.

Mycorrhizae

Water

Minerals

Sugars

Health and Medicine

You might recall that many protists can be harmful to humans and the environment. This is true of fungi as well. A small number of people die every year after eating poisonous mushrooms or spoiled food containing harmful fungi.

You do not have to eat fungi for them to make you sick or uncomfortable. You already read that fungi cause athlete's foot rashes and diaper rashes. Some fungi cause allergies, pneumonia, and thrush. Thrush is a yeast infection that grows in the mouths of infants and people with weak immune systems.

Although fungi can cause disease, scientists also use them to make important medicines. Antibiotics, such as penicillin, are among the valuable medications made from fungi. An accident resulted in the discovery of penicillin. Alexander Fleming was studying bacteria in 1928 when spores of *Penicillium* fungus contaminated his experiment and killed the bacteria. After years of research, this fungus was used to make an antibiotic similar to the penicillin used today. **Figure 16** illustrates how penicillin affects bacterial growth.

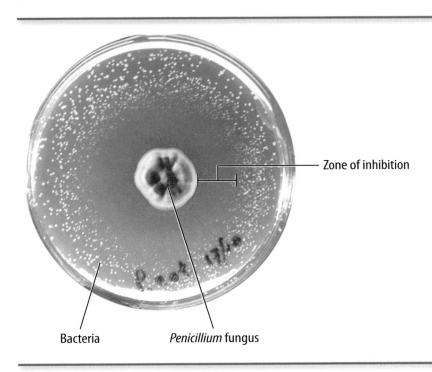

Bacteria

Penicillium fungus

Zone of inhibition

Figure 16 The *Penicillium* fungus that prevents bacteria from growing is used to make penicillin, an antibiotic medicine.

Visual Check How can you tell that the fungi are stopping the bacteria from growing?

Over time, some bacteria have become resistant to the many antibiotics used to fight illness. New antibiotics need to be developed to treat the same diseases. As new species of fungi are discovered and studied, scientists might find new sources of antibiotics and medicines.

 Key Concept Check Describe two ways that fungi are important to humans.

What are lichens?

WORD ORIGIN

lichen
from Greek *leichen*, means
"what eats around itself"

Do you recall the photo at the beginning of the chapter? The structure pictured is a lichen. *A* **lichen** *(LI kun) is a structure formed when fungi and certain other photosynthetic organisms grow together.* Usually, a lichen consists of a sac fungus or club fungus that lives in a partnership with either a green alga or a photosynthetic bacterium. The fungus' hyphae grow in a layer around the algae cells.

Green algae and photosynthetic bacteria are autotrophs, which means they can make their own food using photosynthesis. Lichens are similar to mycorrhizae because both organisms benefit from the partnership. The fungus provides water and minerals while the bacterium or alga provides the sugars and oxygen from photosynthesis.

The Importance of Lichens

Imagine living on a sunny, rocky cliff like the one in **Figure 17.** Not many organisms could live there because there is little to eat. A lichen, however, is well suited to this harsh environment. The fungus can absorb water, help break down rocks, and obtain minerals for the alga or bacterium. They can photosynthesize and make food for the fungus.

Once lichens are established in an area, it becomes a better environment for other organisms. Many animals that live in harsh conditions survive by eating lichens. Plants benefit from lichens because the fungi help break down rocks and create soil. Plants can then grow in the soil, creating a food source for other organisms in the environment.

 Key Concept Check Which two organisms make up a lichen?

Figure 17 Lichens are structures made of photosynthetic organisms and fungi that can live in harsh conditions.

Lichen Structure

Fungal hyphae

Algal cell

Lesson 2 Review

Visual Summary

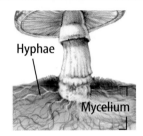

The body of a fungus is made up of thread-like hyphae that weave together to create a network of mycelium.

Hyphae

Mycelium

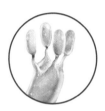

Club fungi produce sexual spores in the basidium.

A lichen is made of fungus and a photosynthetic bacterium or alga.

FOLDABLES

Use your lesson Foldable to review the lesson. Save your Foldable for the project at the end of the chapter.

What do you think NOW?

You first read the statements below at the beginning of the chapter.

4. Mushrooms and yeasts are two types of fungi.

5. Fungi are always helpful to plants.

6. Some fungi can be made into foods or medicines.

Did you change your mind about whether you agree or disagree with the statements? Rewrite any false statements to make them true.

Use Vocabulary

1 **Distinguish** between a basidium and an ascus.

2 **Identify** the structure formed between fungal hyphae and plant roots.

3 **Define** *ascus* in your own words.

Understand Key Concepts

4 **List** the four groups of fungi.

5 **List** the two organisms that make up lichen.

6 Which disease is caused by a fungus?
   A. athlete's foot   C. malaria
   B. influenza   D. pneumonia

Interpret Graphics

7 Review the image below. Describe how this structure helps the fungus survive.

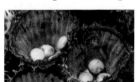

8 **Compare and Contrast** Create a table that compares and contrasts information about sac fungi and zygote fungi.

Sac Fungi	Zygote Fungi

Critical Thinking

9 **Design** a plan for using lichens to convert a harsh cliff environment into a habitat for small plants.

10 **Support** the claim that decomposition is important for the environment.

Math Skills ×÷+   Review

— Math Practice —

11 The number of cells in fungus X doubles every 2 hours. If you begin with 10 cells, how many would be present after 24 hours?

Materials

forceps

lichen sample

paper plate

plastic spoon

slide

dropper

coverslip

microscope

Safety

What does a lichen look like?

Lichens come in a wide variety of textures, colors, sizes, and shapes. A lichen is the partnership between a fungus and another organism—usually an alga but sometimes a photosynthetic bacterium. In this relationship, the alga or the bacterium provides the fungus with food through photosynthesis. The fungus provides the other organism with protection. Under magnification you might be able to see structures belonging to both organisms.

Question

What structures can you see in a lichen?

Procedure

1. Read and complete a lab safety form.

2. With your forceps, break off a tiny piece of lichen and place it on a paper plate. Grind the lichen very gently with the back of a plastic spoon until it is broken into small pieces.

3. Using the spoon, place the ground-up lichen into the well of a slide. Use the dropper to add a few drops of water and then place the coverslip over the well.

4. Observe the lichen under the microscope and make a drawing of your observations in your Science Journal.

5. Label the parts of the lichen you observe. Were you able to see any green algal cells? Where are the hyphae? How is the fungus different in color, shape, and texture?

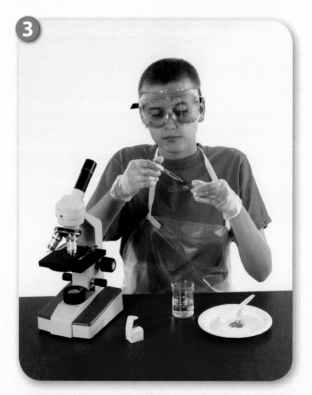

6 Review the structures of the different protists and fungi you have studied so far.

7 Based on your observations from this dissection and your research, determine the ways in which the structures in lichens are similar to and different from those of other organisms.

Analyze and Conclude

8 **Compare** Did you see more fungal structures or algal structures in the slide?

9 **Infer** Based on your observations, do you think a lichen should be classified as a protist, a fungus, neither, or both?

10 **The Big Idea** How do the structures of the algae and fungus benefit each other in a lichen?

Communicate Your Results

Create a poster to represent the data obtained from your investigation. Describe how you used your data to determine the classification of a lichen. Use drawings and photos to support your findings.

 **Extension**

Think of a question about lichen that you might investigate through further observation. Your question might focus on the life cycle of lichen, the best growing conditions, or the lichen as an indicator of air quality. Develop and conduct an experiment to explore your question.

4

Lab Tips

☑ Look for differences in the structures you observe. Try to match the characteristics of these structures to those of protists or fungi.

☑ Begin observing the samples under low magnification first, and then increase as you identify structures.

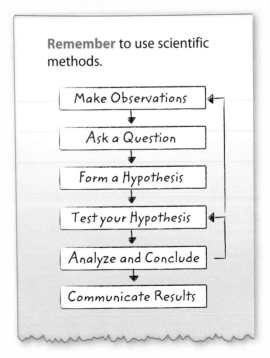

Remember to use scientific methods.

Make Observations
↓
Ask a Question
↓
Form a Hypothesis
↓
Test your Hypothesis
↓
Analyze and Conclude
↓
Communicate Results

Chapter 8 Study Guide

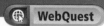

 THE BIG IDEA

Protists and fungi are diverse groups of organisms. They are classified as neither plant nor animal and serve many functions in the ecosystem.

Key Concepts Summary

Lesson 1: What are protists?

- Scientists divide **protists** into three groups based on the type of organisms they most resemble. There are plantlike, animal-like, and funguslike protists.

- Protists are beneficial to humans in many ways. They are used to create many of the useful products you depend on. They also help decompose dead organisms and return nutrients to the environment.

Plantlike	Animal-like	Funguslike

Lesson 2: What are fungi?

- Scientists divide fungi into four groups, based on the type of structures they use for sexual reproduction. The four groups are club fungi, sac fungi, zygote fungi, and imperfect fungi.

- Fungi provide many foods and medicines that people use. In addition, fungi help break down dead organisms and recycle the nutrients into the environment.

- **Lichens** are structures made of a fungus and a photosynthetic organism. Both organisms work together to obtain food, water, and nutrients.

Vocabulary

protist p. 265
algae p. 266
diatom p. 267
protozoan p. 270
cilia p. 270
paramecium p. 270
amoeba p. 271
pseudopod p. 271

hyphae p. 277
mycelium p. 277
basidium p. 278
ascus p. 279
zygosporangia p. 279
mycorrhizae p. 282
lichen p. 284

FOLDABLES® Chapter Project

Assemble your lesson Foldables as shown to make a Chapter Project. Use the project to review what you have learned in this chapter.

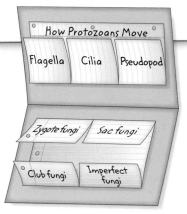

How Protozoans Move

| Flagella | Cilia | Pseudopod |

Zygote fungi | Sac fungi

Club fungi | Imperfect fungi

Use Vocabulary

1 A protist that resembles a tiny animal is called a(n) _____.

2 A fungus and the roots of a plant form a structure called _____ that benefits both organisms.

3 The _____ is a saclike structure on a fungus that produces spores.

4 A(n) _____ is a microscopic, plantlike protist that can resemble glass or gems.

5 Short structures that cover the outside of some protists and help them move are called _____.

6 Fungi grow by extending threadlike body structures called _____.

Link Vocabulary and Key Concepts

◄◎ Concepts in Motion Interactive Concept Map

Copy this concept map, and then use vocabulary terms from the previous page and other terms from this chapter to complete the concept map.

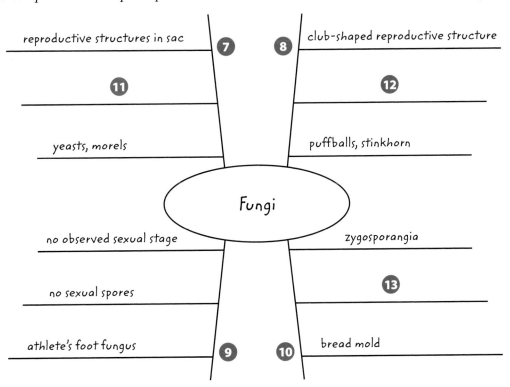

reproductive structures in sac **7** **8** club-shaped reproductive structure

11 **12**

yeasts, morels

puffballs, stinkhorn

Fungi

no observed sexual stage

zygosporangia

no sexual spores

13

athlete's foot fungus **9** **10** bread mold

Understand Key Concepts 🔑

1 Which organism causes red tides when found in large numbers?
A. algae
B. amoebas
C. ciliates
D. diatoms

2 Protists are a diverse group of organisms divided into what three categories?
A. animal-like, plantlike, protozoanlike
B. euglenoid, slime-mold, diatoms
C. plantlike, animal-like, and funguslike
D. green algae, red algae, and kelp

3 Which type of protist is commonly used in ice cream, toothpaste, soups, and body lotions?
A. algae
B. amoebas
C. ciliates
D. diatoms

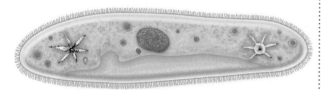

4 The organism in the figure above is a
A. ciliate.
B. diatom.
C. dinoflagellate.
D. kelp.

5 The main function of the hairlike structures surrounding the organism above is
A. decomposition.
B. movement.
C. photosynthesis.
D. reproduction.

6 What type of fungus is bread mold?
A. club
B. imperfect
C. sac
D. zygote

7 Which type of fungus is shown below?

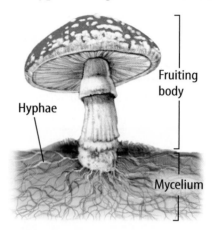

A. club fungus
B. sac fungus
C. zygote fungus
D. imperfect fungus

8 Lichens often consist of
A. plants and animals helping each other.
B. animals and fungi helping each other.
C. protozoans living as parasites on animals.
D. fungi and green algae helping each other.

9 An example of a disease caused by a fungus is
A. athlete's foot.
B. malaria.
C. red tide.
D. the common cold.

10 Sac fungi can be
A. capable of both asexual and sexual reproduction.
B. capable of making their own food.
C. protists.
D. unicellular plants.

11 Which is not a common use of fungi?
A. a predator in forest ecosystems
B. decomposing plant and animal material
C. killing bacteria
D. serving as a food source for other organisms

Critical Thinking

12 **Compare and contrast** plantlike protists with funguslike protists.

13 **Draw** a diagram to show how a parasitic protist can be transferred from one mammal to another and cause malaria. Imagine you are a doctor in an area where malaria is common. How could you prevent the spread of this disease?

14 **Evaluate** Imagine you are asked to justify removing kelp from an area of the ocean. Based on your knowledge of plantlike protists, what benefits or problems would you consider before you decide if the algae should be removed?

15 **Describe** Complete the table below with characteristics of the different types of animal-like protists.

	Number of nuclei	Method of eating	Method of movement
Ciliates			
Flagellates			
Sarcodines			

16 **Explain** how the movement of an amoeba differs from the movement of a dinoflagellate.

17 **List** several products you have used or seen that were made using fungi.

18 **Evaluate** how Alexander Fleming's experiments helped determine the importance of fungi to medicine.

Writing in Science

19 **Design** a brochure for a tour in which people could see several different types of lichens and fungi. What locations would be included and which organisms would people be likely to observe?

REVIEW THE B|G IDEA

20 Explain how decomposers such as protists and fungi play an important role in the environment.

21 What is the organism shown below and how does it affect the environment?

Math Skills ×÷+

Review

—— Math Practice ——

Calculating Growth

22 The number of cells of Fungus Q doubles every three hours. If you begin with 1,000 cells, how many will there be after 12 hours?

23 Scientists want to know if an antibiotic is effective in treating a fungal infection. They start with two colonies of 100 cells each. The table shows what happens during the first two days.

	Day 2 Number of Cells	Day 3 Number of Cells
Untreated fungus	400	1600
Antibiotic A	200	300

a. How long does it take the untreated fungus to double?

b. What effect does the antibiotic have on the growth rate of the fungus?

Standardized Test Practice

Record your answers on the answer sheet provided by your teacher or on a sheet of paper.

Multiple Choice

1 Which often live on decaying leaves in a forest?

  A diatoms

  B dinoflagellates

  C sarcodines

  D slime molds

Use the diagram below to answer question 2.

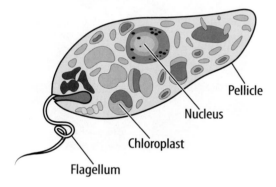

2 This organism has plant and animal characteristics. Which characteristic makes the organism plantlike?

  A chloroplast

  B flagellum

  C nucleus

  D pellicle

3 Which statement is false?

  A Fungi cause bread to rise.

  B Fungi cause dead organisms to decay.

  C Fungi are sources of antibiotics.

  D Fungi use sunlight and produce food.

4 Which protists have cell walls that look like glass?

  A algae

  B diatoms

  C dinoflagellates

  D euglenoids

Use the diagram below to answer question 5.

Mycorrhizae

5 What does the plant give the fungus that surrounds its roots?

  A antibiotics

  B minerals

  C sugars

  D water

Use the diagram below to answer question 6.

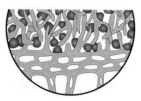

Lichen

6 Which organisms, combined with green algae, form this structure?

  A bacteria

  B fungi

  C plants

  D protozoans

7 Suppose a pond contains no living or decaying organisms. Which could be added to the pond to act as producers?

   A algae

   B ciliates

   C sarcodines

   D water molds

Use the diagram below to answer questions 8 and 9.

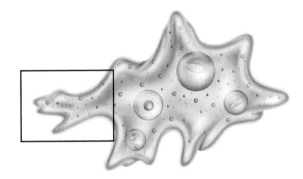

8 What is the function of the boxed area of this microscopic organism?

   A cellular respiration

   B defense

   C locomotion

   D photosynthesis

9 To which group does this organism belong?

   A animal-like protists

   B animals

   C fungi

   D funguslike protists

Constructed Response

Use the diagram below to answer question 10.

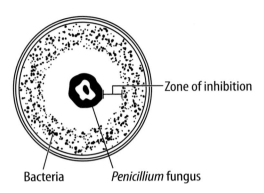

Zone of inhibition

Bacteria   *Penicillium* fungus

10 What is the interaction between the fungus and the bacteria?

11 Copy and complete the table below.

Group of Protists	Beneficial Effects	Harmful Effects
Plantlike		
Animal-like		
Funguslike		

12 For one of the groups of protists you listed in the table in question 11, explain why their beneficial effects are important to other organisms.

13 Identify two ways protists might reproduce. Describe the offspring that result from each method. What is an advantage of each method?

NEED EXTRA HELP?													
If You Missed Question...	1	2	3	4	5	6	7	8	9	10	11	12	13
Go to Lesson...	1	1	2	1	2	2	1	1	1	2	1,2	1,2	1

Plant Diversity

THE BIG IDEA Why are plants in so many different environments on Earth?

Inquiry Why so many plants?

There are many different kinds of plants growing here! What plant characteristics do you think enable such a wide variety of plants to grow in one place?

- Where else do plants grow?

- How are those plants different from the plants growing in this environment? How are they similar?

Get Ready to Read

What do you think?

Before you read, decide if you agree or disagree with each of these statements. As you read this chapter, see if you change your mind about any of the statements.

1 All plants produce flowers and seeds.

2 Humans depend on plants for their survival.

3 Some plants move water only by diffusion.

4 Mosses can grow only in moist, shady places.

5 Some mosses and gymnosperms are used for commercial purposes.

6 All plants grow, flower, and produce seeds in one growing season.

ConnectED Your one-stop online resource

connectED.mcgraw-hill.com

Video   WebQuest

Audio   Assessment

Review   Concepts in Motion

Inquiry   Multilingual eGlossary

Reading Guide

Key Concepts

ESSENTIAL QUESTIONS

- What characteristics are common to all plants?

- What adaptations have enabled plant species to survive Earth's changing environments?

- How are plants classified?

Vocabulary

producer p. 298

cuticle p. 299

cellulose p. 299

vascular tissue p. 300

g Multilingual eGlossary

What is a plant?

Inquiry Why So Successful?

This plant cell has parts animal cells don't have. How do you think those cellular parts help plants live? What other parts do plants have that enable them to be successful in so many diverse environments on Earth?

Launch Lab

What is a plant?

A plant often is described as a living thing that makes its own food, has leaves and stems, and is green in color. Even with the many different types of plants on Earth, this description often holds true. Or does it?

1. Examine the photos in the table below.

1	2	3	4	5	6

2. Number from 1 to 6 in your Science Journal. Next to each number write *yes* if you think the photo is of a plant or *no* if you think the photo is not of a plant.

Think About This

1. Which photos do you think are plants? Explain your choices.

2. Visualize each object without the background, and decide if you want to make changes in your list.

3. **Key Concept** What characteristics are common to each plant in these pictures?

Characteristics of Plants

You might not think about plants often, but they are an important part of life on Earth. As you read this lesson, look for the characteristics that make plants so important to other organisms.

Cell Structure

Plants are made of eukaryotic cells. Recall that eukaryotic cells have membrane-bound organelles. Some of a plant cell's organelles are shown in **Figure 1.** A plant cell differs from an animal cell because it contains chloroplasts and a cell wall. Chloroplasts convert light energy to chemical energy. The cell wall provides support and protection. A mature plant cell also has one or two vacuoles that store a watery liquid called sap.

Reading Check Describe the structure of a plant cell.

Concepts in Motion Animation

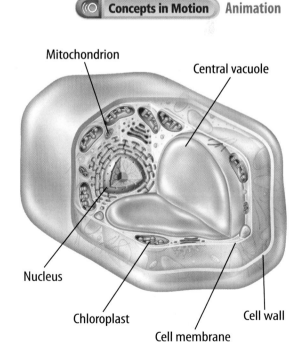

Mitochondrion

Central vacuole

Nucleus

Chloroplast

Cell membrane

Cell wall

Figure 1 Plant cells and animal cells have many of the same organelles.

 Figure 2 Some plants, such as the reproductive stage of some ferns, are microscopic. Other plants, such as redwood trees, are huge.

WORD ORIGIN ·

producer
from Latin *producer*, means "lead or bring forth, draw out"

· ·

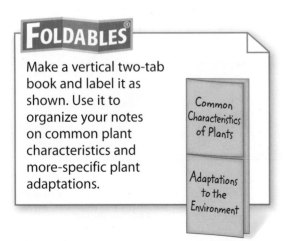

FOLDABLES®

Make a vertical two-tab book and label it as shown. Use it to organize your notes on common plant characteristics and more-specific plant adaptations.

Common Characteristics of Plants

Adaptations to the Environment

Multicellular

Plants are multicellular. This means they are made of many cells. The cells carry out specialized functions and work together to keep the plant alive. As shown in **Figure 2,** some plants are microscopic while others are some of the largest organisms on Earth.

Producers

Organisms that use an outside energy source, such as the Sun, to make their own food are called **producers.** Plants are producers. They make their own food, a simple sugar called glucose, during a process called photosynthesis. All other organisms rely on producers, either directly or indirectly, for their sources of food.

Key Concept Check What characteristics are common to all plants?

Plant Adaptations

Millions of years ago, there were no land plants. Scientists hypothesize that present-day land plants and green algae evolved from a common ancestor. They base their hypothesis on chemical similarities between green algae and plants. Some of the pigments in green algae and land plants are the same kind. There also are DNA similarities between these two groups of organisms.

The first land plants probably lived in moist areas. Life on land would have provided some advantages to plants. There would have been plenty of sunlight for photosynthesis to occur. The air that surrounded those plants would have been a mixture of gases, including carbon dioxide—another thing needed for photosynthesis. As land plants became more abundant, the amount of oxygen in the atmosphere increased because oxygen is a product of photosynthesis.

Plant species also had to adapt to survive without being surrounded by water. Many of the characteristics we now see in plants are adaptations to life on land.

Protection

One advantage to life on land is a constant supply of air that contains carbon dioxide. As you just read, carbon dioxide is needed for photosynthesis. Many plants have a *waxy, protective layer on their leaves, stems, and flowers called the* **cuticle.** It is made of a waxy substance that is secreted by the cells. Its waxy nature slows the evaporation of water from a plant's surface. This covering also provides some protection from insects that might harm a plant's tissues.

 Reading Check How does a plant's cuticle protect it?

Support

The water that surrounds aquatic plants supports them. Land plants must provide their own support. Like all cells, a plant cell has a cell membrane. Recall that a rigid cell wall surrounds the cell membrane in a plant cell. The cell wall provides support and is made of cellulose. **Cellulose** *is an organic compound made of chains of glucose molecules.* Many land plants also produce a chemical compound called lignin (LIG nun). Lignin strengthens cellulose and makes it more rigid. The piece of wood shown in **Figure 3** is mostly made of cellulose and lignin.

Figure 3 🔑 The combined strength of all of a plant's cell walls provides support for the plant.

Inquiry MiniLab — 30 minutes

How does water loss from a leaf relate to the thickness of the cuticle?

Plants need water for structural support and nutrients. Land plants developed a thickened layer of waxy material on their leaves, called a cuticle, that prevented water loss. The thickness of the cuticle varies from one plant type to another.

1. Read and complete a lab safety form.
2. In your Science Journal, make a table like the one below.

Leaf	Mass (mg) Day 1	Mass (mg) Day 2	Difference in Mass	% Decrease
1				
2				
3				
4				

3. Choose **leaves from four different plants.** Use a **balance** to determine the mass, in milligrams, of each leaf. Record the data in your table.
4. Lay each leaf on a **small tray,** making sure that none of the leaves touch other leaves. Leave them undisturbed overnight.
5. Measure the mass of each leaf again the following day. Record the data.
6. Calculate the difference in mass from day 1 to day 2. Calculate the percent difference.

Analyze and Conclude

1. **Analyze** Did you observe any changes in the initial and final masses of the leaves? Explain.
2. **Compare** the thickness of the leaves and determine which type(s) lost the least amount of water.
3. 🔑 **Key Concept** Consider the structure of the leaf that lost the least amount of water. Why do you think this adaptation is helpful?

Transporting Materials

In order for a plant to survive, water and nutrients must move throughout its tissues. In some plants such as mosses, these materials can move from cell to cell by the processes of osmosis and diffusion. This means that water and other materials dissolved in water move from areas of a plant where they are more concentrated to areas where they are less concentrated. However, other plants such as grasses and trees have specialized tissues called vascular tissue. **Vascular tissue** *is composed of tubelike cells that* transport *water and nutrients in some plants.* Vascular tissue can carry materials throughout a plant—great distances if necessary, up to hundreds of meters. You will read more about vascular tissues in Lesson 3 of this chapter.

ACADEMIC VOCABULARY

transport
(verb) to carry somebody or something

Reproduction

Water carries the reproductive cells of aquatic plants from plant to plant. How do you think land plants reproduce without water? Land plants evolved other strategies for reproduction. Some plants have water-resistant seeds or spores that are part of their reproductive process. Seeds and spores move throughout environments in different ways. These include animals and environmental factors such as wind and water. Several methods of seed dispersal are shown in **Figure 4.**

 Key Concept Check What adaptations of plants have enabled them to survive Earth's changing environments?

Adaptations for Seed Dispersal

Figure 4 Plants have developed different methods of dispersing their seeds.

Coconut seeds float in water.

Wind carries milkweed seeds.

Burrs, which contain seeds, cling to clothes, fur, or feathers.

Liverwort

Pine Tree

Magnolia

Moss

Plant Classification

You might recall that kingdoms such as the animal kingdom consist of smaller groups called phyla. Members of the plant kingdom are organized into groups called divisions instead of phyla. Like all organisms, each plant has a two-word scientific name. For example, the scientific name for a red oak is *Quercus rubra*.

 Key Concept Check How are plants classified?

Seedless Plants

Liverworts and mosses, such as the ones shown in **Figure 5,** reproduce by structures called spores. Plants that reproduce by spores often are called seedless plants. Seedless plants do not have flowers. Some seedless plants do not have vascular tissue and are called nonvascular plants. Others, such as ferns, have vascular tissue and are called vascular plants. Seedless plants are classified into several divisions.

Seed Plants

Most of the plants you see around you, such as pine trees, grasses, petunias, and oak trees, are seed plants. Almost all the plants we use for food are seed plants. Some seed plants have flowers that produce fruit with one or more seeds. Others, such as pine trees, produce their seeds in cones. Each seed has tissues that surround, nourish, and protect the tiny plant embryo inside it. It is thought that all present-day plants originated from a common ancestor, an ancient green algae, as shown in **Figure 6** on the next two pages.

Figure 5 Liverworts and mosses reproduce by producing spores. Both pine and magnolia trees produce seeds.

✅ **Visual Check** How are the pine tree and the magnolia tree alike? How are they different?

Ferns are vascular plants, but fern reproduction includes spores.

Hornworts often grow in fields or along roads.

Mosses are the most abundant nonvascular plants. Mosses generally grow in shady, moist places.

Club mosses are spore-producing vascular plants.

Nonvascular Plants

Vascular Plants

Liverworts grow in almost every habitat on Earth.

Ancient green algae are thought to be the ancestors of land plants as well as present-day green algae.

Figure 6 This tree represents the evolutionary relationships among plant divisions.

✔ **Visual Check** What type of plant is thought to be the ancestor of land plants?

One stage of this plant's life cycle reminded people of a horse's tail. **Horsetails** also have been called scouring rushes due to their abrasive texture.

Conifers are the largest and most diverse division of gymnosperms.

There is only one species of **ginkgo** alive today. It is often used as an ornamental tree.

Seedless Plants

Gymnosperms

Cycads usually grow in tropical regions.

Seed Plants

Angiosperms

Tulips grow best in climates that have long, cool springs.

Grass flowers typically are small and easily overlooked.

Visual Summary

Plants are multicellular producers.

Water carries the reproductive cells of aquatic plants from plant to plant. Land plants evolved different reproductive strategies to ensure their survival.

Members of the plant kingdom are classified into groups called divisions.

FOLDABLES

Use your lesson Foldable to review the lesson. Save your Foldable for the project at the end of the chapter.

What do you think NOW?

You first read the statements below at the beginning of the chapter.

1. All plants produce flowers and seeds.

2. Humans depend on plants for their survival.

Did you change your mind about whether you agree or disagree with the statements? Rewrite any false statements to make them true.

Use Vocabulary

1. **Write** a sentence using the term *vascular tissue*.

2. **Define** *cellulose* in your own words.

Understand Key Concepts

3. Which structure helps support plant cells?
   - **A.** cell wall
   - **B.** chloroplast
   - **C.** mitochondria
   - **D.** ribosome

4. **List** the common characteristics of plants.

5. **Describe** an example of a plant adaptation that helps plants survive on land.

6. **Distinguish** between seedless and seed plants.

Interpret Graphics

7. **Examine** the diagram below and list the cell structures that identify it as a plant cell.

8. **Organize** Copy and fill in the graphic organizer below. In the center oval write *Adaptations to Life on Land*. Fill in the other ovals with plant adaptations.

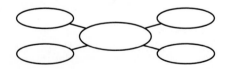

Critical Thinking

9. **Hypothesize** which of the adaptations to life on land might be missing in a plant that grows in shallow water.

10. **Assess** the importance of plants as producers.

Trees in the Sea

How can a plant survive in salt water?

Mangrove forests thrive along warm coastlines. Birds perch in the treetops, while young fish and other animals hide among the widespread underwater roots. These amazing trees can survive where others cannot—in salt water. Salt water kills most plants. It causes freshwater to diffuse out of plant cells, and they collapse. In addition, most plants would not survive with their roots under water because they cannot absorb oxygen from water. But mangrove trees have special traits that help them survive along the ocean's edge.

AMERICAN MUSEUM OF NATURAL HISTORY

1 Leaves
- A mangrove's thick leaves store freshwater. The waxy coating, or cuticle, covering the leaves reduces the amount of water lost by evaporation.
- Some mangroves release salt through glands in their leaves. When the leaves drop off the plant, the salt goes with them.

2 Seedlings
- Unlike most other plant seeds that sprout in soil, mangrove seeds begin sprouting on the tree. They're ready to take root as soon as they drop.
- A mangrove seedling can survive floating in seawater for up to a year until it finds a suitable environment in which to grow.

3 Bark
- Some mangrove species store salt in bark that later peels away.
- Some mangrove bark also has tiny holes. Oxygen can enter a mangrove through these holes.

4 Roots
- Mangrove roots filter salt out of salt water.
- The roots arch over the water and absorb oxygen through tiny pores. These pores close when salt water covers the roots during a high tide.

It's Your Turn

DIAGRAM Choose a species of tree in your area. Research the traits of that species, such as bark, roots, leaves, or seeds, that help it be successful in its environment. Draw and label a diagram of the tree that illustrates your findings.

Reading Guide

Key Concepts
ESSENTIAL QUESTIONS

How are nonvascular and vascular seedless plants alike, and how are they different?

Vocabulary
rhizoid p. 307
frond p. 309

g Multilingual eGlossary

Seedless Plants

1.0 mm

Inquiry **Spores Instead of Seeds?**

Ferns are vascular seedless plants. They reproduce using spores instead of seeds. This type of fern produces spores in clusters on one side of its fronds. What characteristics do all vascular seedless plants share? How do vascular seedless plants compare to nonvascular seedless plants?

Which holds more water?

Peat moss is the common name of approximately 300 different types of mosses. The partially decomposed remains of these moss plants form peat moss. Some potting soils contain peat moss.

1. Read and complete a lab safety form.
2. Fill a **250-mL beaker** with **potting soil** to about 3 cm from the top.
3. In a **tub,** mix equal parts **peat moss** and potting soil. Fill a second 250-mL beaker to the same level with this mixture.
4. Pour 30 mL of water into each beaker. Examine the beakers after 5 minutes and record your observations in your Science Journal.
5. After another 5 minutes, place each beaker on its side in an **aluminum pie pan.** Record your observations.

Think About This

1. How quickly did each soil mixture absorb the water?

2. What happened when you placed the beakers on their sides in the pie pans?

3. 🔑 **Key Concept** Why do you think peat moss is added to potting soil? How might this benefit plants?

Nonvascular Seedless Plants

If someone asked you to make a list of plants, your list might include plants such as your favorite flowers or trees that grow near your home. You probably would not include any nonvascular seedless plants on your list.

Many scientists refer to all nonvascular seedless plants as bryophytes (BRI uh fites). These plants usually are small. Because they lack tubelike structures, called vascular tissue, that transport water and nutrients, the bryophytes usually live in moist environments. Materials move from cell to cell by diffusion and **osmosis.**

Because bryophytes do not have vascular tissue, they do not have roots, stems, or leaves. They have rootlike structures called rhizoids, shown in **Figure 7. Rhizoids** *are structures that anchor a nonvascular seedless plant to a surface.* Rhizoids can be unicellular—consisting of only one cell—or they can be multicellular. The photosynthetic tissue of bryophytes is often only one cell layer thick. This layer does not have a cuticle, which most other plants have. Reproduction is by spores and requires water. Mosses, liverworts, and hornworts are bryophytes.

✓ **Reading Check** What characteristics are common in bryophytes?

REVIEW VOCABULARY

osmosis
the diffusion of water molecules

Figure 7 🔑 Rhizoids anchor bryophytes to a surface, such as soil, rocks, and the bark of trees.

✓ **Visual Check** How is the structure of rhizoids well-suited to the function of rhizoids?

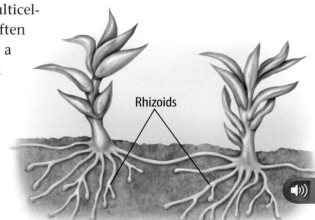

Rhizoids

Mosses

You might be familiar with the most common bryophytes—the mosses. These small, green plants grow in forests, in parks, and sometimes even in the cracks of sidewalks. Mosses usually grow in shady, damp environments, but they are able to survive periods of dryness. As shown in **Figure 8,** mosses have leaf-like structures that grow on a stemlike structure called a stalk. They have multicellular rhizoids.

Mosses play an important role in the ecosystem. They are often the first plants to grow in barren areas or after a natural disturbance such as a fire or a mudslide. The ability of mosses to retain large amounts of water makes peat moss a useful additive for potting soil, as you discovered in the Launch Lab. This moss has been used to enrich soil and as a heating source.

Liverworts

Hundreds of years ago, people thought that this plant could be used to treat liver diseases. Liverwort also gets its name from its appearance—it resembles the flattened lobes of a liver. The rhizoids of liverworts are unicellular. The two common forms of liverworts are leafy and thallose (THA los) liverworts, as shown in **Figure 8.**

Hornworts

The long, hornlike reproductive structures shown in **Figure 8** give this group of plants its name. These reproductive structures produce spores. Hornworts are only about 2.5 cm in diameter. One unusual characteristic of hornworts is that each of its photosynthetic cells has only one chloroplast.

Figure 8 Mosses, liverworts, and hornworts all lack vascular tissue.

Nonvascular Seedless Plants 🔑

The green, leafy structures of mosses are not considered leaves because they do not contain vascular tissue.

A leafy liverwort, on the left, also lacks vascular tissue. The lobes of a thallose liverwort, on the right, can be as thin as one cell layer thick.

Until hornworts produce their reproductive structures, or "horns," they can easily be mistaken for liverworts.

Vascular Seedless Plants

Over 90 percent of plant species are vascular plants. Unlike nonvascular plants, they contain vascular tissue in their stems, roots, and leaves. Because vascular plants contain tubelike structures that transport water and nutrients, these plants generally are larger than nonvascular plants. However, present-day vascular seedless plants are smaller than their ancient ancestors. These ancient plants grew as tall as trees. Much of the fossil fuels that we use today came from the remains of these ancient plants.

Ferns

The **fronds,** *or leaves of ferns,* make up most of a fern. Ferns range in size from a few centimeters to several meters tall, such as the one in **Figure 9.** They grow in a variety of habitats, including damp, swampy areas and dry, rocky cliffs. Ferns often are houseplants or grow in gardens. Some people consider young fronds, also called fiddleheads, a gourmet treat.

Club Mosses

Unlike mosses, club mosses have roots, stems, and leaves. Club mosses, shown in **Figure 10,** are small plants that rarely grow taller than 50 cm. The stems often grow along the ground. The leaves are scalelike. The spores of club mosses make a fine powder that is so flammable that it has been used to make fireworks!

Horsetails

As shown in **Figure 11,** horsetails have small leaves growing in circles around the stems. Horsetail stems are hollow, and the tissues contain silica, a mineral in sand, that makes them abrasive. They once were used for scrubbing pots. Horsetails can be grown in water gardens but tend to spread rapidly.

 Key Concept Check How are nonvascular and vascular seedless plants alike? How are they different?

▲ **Figure 9** Tree ferns, such as this one, once were the dominant plants on Earth.

▲ **Figure 10** Club mosses reproduce by producing spores in two or three cylindrical, yellow-green colored cones.

▲ **Figure 11** The hollow stem of a horsetail is its main photosynthetic structure.

Lesson 2 Review

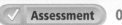
Visual Summary

Many scientists refer to all nonvascular seedless plants as bryophytes.

Because vascular plants contain tube-like structures that transport water and nutrients, these plants usually are larger than nonvascular plants.

Humans use both vascular and nonvascular plants for many purposes.

FOLDABLES

Use your lesson Foldable to review the lesson. Save your Foldable for the project at the end of the chapter.

What do you think NOW?

You first read the statements below at the beginning of the chapter.

3. Some plants move water only by diffusion.

4. Mosses can grow only in moist, shady places.

Did you change your mind about whether you agree or disagree with the statements? Rewrite any false statements to make them true.

Use Vocabulary

1 **Write** a sentence using the term *rhizoid*.

Understand Key Concepts 🔑

2 Which does NOT belong with the others?
  A. club moss   C. horsetail
  B. fern   D. liverwort

3 **List** the different types of bryophytes.

Interpret Graphics

4 **Examine** the picture of the club moss below, and explain why some people call club mosses "ground pine."

5 **Organize** Copy and fill in the table below to summarize the different types of vascular and nonvascular seedless plants.

Types of Seedless Plants	
Vascular	Nonvascular

Critical Thinking

6 **Predict** how well moss plants would grow in the desert. Explain your reasoning.

7 **Compare and contrast** nonvascular and vascular seedless plants.

How do differences in plant structures reflect their environments?

Plants with the same structures can appear to be very different from each other. Often the same structures serve different purposes, depending on the environment in which the plant lives. You can compare and contrast the same structures on two different plants to learn about their needs in different environments.

Materials

cactus plant

rain forest plant

Safety

Learn It

Comparing and contrasting allows you to learn more about things than observing them separately. Noting the similarities and differences of these plants can help you learn which features are adaptations to a specific environment.

Try It

1 Read and complete a lab safety form.

2 Observe a cactus plant and a rain forest plant in your classroom.

⚠ Take care touching and handling the cactus plant.

3 Compare and contrast the leaves of both plants. Describe the leaves of each plant in a table. Hint: The spines of the cactus plant are modified leaves, while the green, thick parts are stems.

4 Observe the pictures of the desert environment and the rain forest environment. Think about how the leaves of both plants might have a different function in each environment. Fill out the next row of the table with your thoughts and descriptions.

5 Compare and contrast the stems. Describe the stems of each plant in your table.

6 Think about how the stems of both plants might have a different function in each environment. Fill out the next row of the table with your thoughts and descriptions.

Rain forest

Desert

Apply It

7 **Describe** which parts of each plant capture energy from sunlight.

8 **Analyze** how the availability of sunlight in each environment might affect the size and number of light-capturing structures on both plants.

9 **Infer** how both plants store water differently based on the differences in their structures.

10 🗝 **Key Concept** Which plant structures were similar in both environments? Explain why these structures were so similar.

Lesson 3

Reading Guide

Key Concepts

ESSENTIAL QUESTIONS

- What characteristics are common to seed plants?
- How do other organisms depend on seed plants?
- How are gymnosperms and angiosperms alike, and how are they different?
- What adaptations of flowering plants enable them to survive in diverse environments?

Vocabulary

cambium p. 314

xylem p. 314

phloem p. 315

stoma p. 317

g Multilingual eGlossary

▢ Video BrainPOP®

Seed Plants

Inquiry A Great Relationship?

The berry the bird is eating has seeds in it. The bird depends on the tree for food—the berries. The tree helps the bird live. How do you think the bird is helping the tree live and be successful in its environment? What other adaptations might help seed plants be successful in different environments?

What characteristics do seeds have in common?

Seed plants have two characteristics in common: they have vascular tissue and seeds for reproduction. Do seeds also have common characteristics?

1. Read and complete a lab safety form.

2. Examine the **several types of seeds** on your **tray.**

3. In your Science Journal, make a grid of 2-cm by 2-cm squares to classify the seeds. Choose your own criteria to classify the seeds; for example, color, texture, size, or other special characteristics.

4. Label the columns and rows of your grid with the characteristics you chose.

5. Place the seeds on the section(s) of your grid where they belong based on the criteria you have chosen to classify them.

Think About This

1. Explain why you placed some of the seed samples in the same square.

2. Could some of the seeds have been placed in more than one square? Elaborate.

3.  **Key Concept** Do any of the seeds share common characteristics? Explain.

Characteristics of Seed Plants

Have you ever eaten corn, beans, peanuts, peas, or pine nuts, such as the ones shown in **Figure 12?** They are all examples of edible seeds. Recall that a seed contains a tiny plant embryo and nutrition for the embryo to begin growing. There are more than 300,000 species of seed plants on Earth.

Seed plants are organized into two groups—cone-bearing seed plants, or gymnosperms (JIHM nuh spurmz) and flowering seed plants, or angiosperms (AN gee uh spurmz). All seed plants have vascular tissue that transports water and nutrients throughout the plant. This means they also have roots, stems, and leaves. You will read more about the characteristics of seed plants in this lesson.

 Key Concept Check What characteristics do all seed plants have in common?

Corn kernels

Peas

Pine nuts

Peanuts

Beans

Figure 12 Of the major plant parts, seeds are the most important source of human food.

SCIENCE USE V. COMMON USE ....

tissue
Science Use a group of cells in an organism

Common Use a piece of absorbent paper

WORD ORIGIN ....

xylem
from Greek *xylon*, means "wood"

Vascular Tissue

All seed plants contain vascular **tissues** in their roots, stems, and leaves. This tissue transports water and nutrients throughout a plant. The two types of vascular tissue are xylem (ZI lum) and phloem (FLOH em). *The* **cambium** *is a layer of tissue that produces new vascular tissue and grows between xylem and phloem.* How do xylem and phloem differ? Keep reading to find out.

 Reading Check What are the two types of vascular tissue? In what structures of a vascular plant would such tissues be?

Xylem *One type of vascular tissue—***xylem***—carries water and dissolved nutrients from the roots to the stem and the leaves.* Due to the thickened cell walls of some xylem cells, this tissue also provides support for a plant.

Two kinds of xylem cells are tracheids (TRAY kee udz) and vessel elements. All vascular plants have xylem that is composed of tracheids. As shown in **Figure 13,** tracheid cells are long and narrow with tapered ends. The tracheid cells grow end-to-end and form a strawlike tube. Water can flow from one cell to another, passing through openings or pits in the end wall of each cell. Tracheid cells die at maturity, leaving a hollow tube. This enables water to flow freely through them.

In addition to tracheids, xylem in flowering plants includes a type of cell called a vessel element. The diameter of a vessel element is greater than that of a tracheid. The end walls of vessel elements have large openings where water can pass through, as shown in **Figure 13.** In some vessel elements, the end walls are completely open. Vessel elements are more efficient at transporting water than tracheids are.

 Reading Check Why can water and dissolved nutrients flow so freely through xylem?

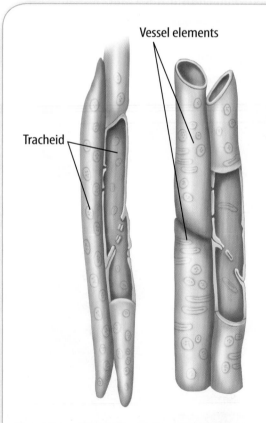

Vessel elements

Tracheid

Figure 13 🔑 The cell walls between individual vessel elements have larger openings than the cell walls between tracheids.

Phloem *Another type of vascular tissue—***phloem***—carries dissolved sugars throughout a plant.* It is composed of two types of cells—sieve-tube elements and companion cells.

Sieve-tube elements are specialized phloem cells. These long, thin cells are stacked end-to-end and form long tubes. The end walls have holes in them, as shown in **Figure 14.** The cytoplasm of a sieve-tube element lacks many organelles, including a nucleus, mitochondria, and ribosomes.

Each sieve-tube element has a companion cell next to it that contains a nucleus. A companion cell helps control the functions of the sieve-tube element.

✓ **Reading Check** List and describe the function of each of the two types of vascular tissue in plants.

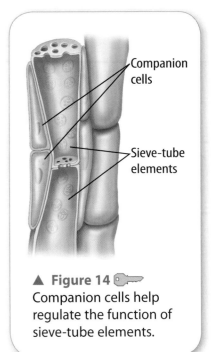

Companion cells

Sieve-tube elements

▲ **Figure 14** 🔑 Companion cells help regulate the function of sieve-tube elements.

Roots

Even though the roots of most plants are never seen, they are vital to a plant's survival. Roots anchor a plant, either in soil or onto another plant or an object such as a rock. All roots help a plant stay upright. Some plants have roots that spread out in all directions several meters from a plant's stem.

All root systems, such as the one shown in **Figure 15,** help a plant absorb water and other substances from the soil.

Plants such as radishes and carrots store food in their roots. This food can be used to grow new plant tissues after a dry period or a cold season. Sugar stored in the roots of sugar maple trees over the winter is converted to maple sap in the spring. Farmers drain some of the sap from these trees and boil it to make maple syrup.

Figure 15 🔑 Roots help anchor a plant and absorb water and minerals from the soil. ▼

Figure 16 🔑 An herbaceous stem supports the lilies growing at the base of the tree. A woody stem supports this tree.

Stems

The part of a plant that connects its roots to its leaves is the stem. In plants such as the tree in **Figure 16,** the stem is obvious. Other plants, such as the potato and the iris, have underground stems that are often mistaken for roots.

Stems support branches and leaves. Their vascular tissues transport water, minerals, and food. Xylem carries water and minerals from the roots to the leaves. The sugar produced during photosynthesis flows through a stem's phloem to all parts of a plant. Another important function of stems is the production of new cells for growth, but only certain regions of a stem produce new cells.

Plant stems usually are classified as either herbaceous or woody. Woody stems, such as the one shown in **Figure 16,** are stiff and typically not green. Trees and shrubs have woody stems. Herbaceous stems usually are soft and green, as also shown in **Figure 16.**

Reading Check What is the importance of a stem to a plant?

Inquiry MiniLab

25 minutes

How can you determine the stems, roots, and leaves of plants?

When the environment changes, the functions of a plant's roots, stems, and leaves do not change. Instead, the structure of one or more of these parts will change in the species over time and adapt to the change in the environment.

1. Read and complete a lab safety form.

2. Examine the **four plants** on your **lab tray.** Gently remove a root, a stem, and a leaf from each plant.

3. Classify the roots into groups based on their characteristics. Do the same with the stems and the leaves. Record your classifications in your Science Journal.

4. In your Science Journal, describe some characteristics of roots, stems, and leaves that you observed in each of the plant parts.

Analyze and Conclude

1. **Describe** how you classified the roots, the stems, and the leaves.

2. **Analyze** any unusual characteristics you observed in any of the plant parts that might be an adaptation. Explain your reasoning.

3. 🔑 **Key Concept** What characteristics were common to all the plants you examined?

Leaves

Leaves come in many shapes and sizes. Most leaves have an important function in common—they are the major site of photosynthesis for the plant. By capturing light energy and converting it to chemical energy, leaves provide the plant's food.

As shown in **Figure 17,** most leaves are made of layers of cells. The top and bottom layers of a leaf are made of epidermal (eh puh DUR mul) tissue. Epidermal cell walls are transparent, and light passes through them easily. These cells produce a waxy outer layer called the cuticle. The cuticle helps reduce the amount of water that evaporates from a leaf. *Most leaves have small openings in the epidermis called* **stomata** (STOH muh tuh; singular, stoma). When the stomata open, carbon dioxide, oxygen, and water vapor can pass through them. Two guard cells surround each stoma and control its size.

Below the upper epidermis are rows of tightly packed cells called palisade (pa luh SAYD) mesophyll (MEH zuh fil) cells. Photosynthesis mainly occurs in these cells. Under the palisade mesophyll cells is the spongy mesophyll layer. The arrangement of these cells enables gases to diffuse throughout a leaf. A leaf's xylem and phloem transport materials throughout the leaf.

Angiosperm and gymnosperm leaves each have some unique characteristics. An angiosperm leaf tends to be flat with a broad surface area. A gymnosperm leaf is usually needlelike or scalelike and often has a thick cuticle. Because gymnosperms often grow in drier areas, these characteristics help conserve water.

Figure 17 The structure of a leaf is well suited to its function of photosynthesis.

Visual Check Describe the location and appearance of the stomata. What role do they play in photosynthesis?

Leaf Anatomy 🔑

(◎ Concepts in Motion) Animation

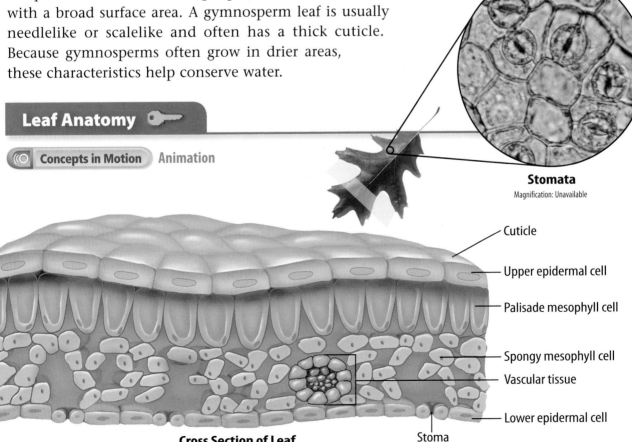

Stomata
Magnification: Unavailable

Cuticle

Upper epidermal cell

Palisade mesophyll cell

Spongy mesophyll cell

Vascular tissue

Lower epidermal cell

Cross Section of Leaf　　Stoma

Types of Gymnosperms

▲ Gnetophyte

Conifer ▶

Cycad ▼

◀ Ginkgo

Figure 18 Gymnosperms are a diverse group of plants, and all produce seeds without surrounding fruits.

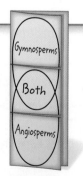

FOLDABLES®

Make a three-tab Venn book, and label it as shown. Use it to compare and contrast seed plants.

Gymnosperms / Both / Angiosperms

Gymnosperms

In gymnosperms, seeds are produced in a cone. This group includes the oldest plant (the bristlecone pine at 4,900 years old), the tallest plant (the coast redwood that can grow to 115 m), and perhaps Earth's largest organism (the sequoia). Conifers, such as spruces, pines, and redwoods, might be the most familiar gymnosperms to you. However, there are several other types of gymnosperms, as shown in **Figure 18.**

Conifers grow on all the world's continents except Antarctica. Cycads usually grow in tropical regions. Although cycads might resemble ferns, they are seed plants. DNA evidence indicates that they are closely related to other gymnosperms. One gymnosperm group has only one species—ginkgo. Ginkgoes have broad leaves and are popular as an ornamental tree in urban areas. The gnetophytes (NEE tuh fites), another type of cone-bearing plant, are an unusual and diverse group of gymnosperms, as shown in **Figure 18.**

Humans rely on gymnosperms for a variety of uses, including building materials; paper production; medicines; and as ornamental plants in gardens, along streets, and in parks.

318 Chapter 9

EXPLAIN

Angiosperms

There are more than 260,000 species of flowering plants, or angiosperms. Angiosperms began to flourish about 80 million years ago. They grow in a variety of habitats, from deserts to the tundra. Anytime you have stopped to smell a flower, you have enjoyed an angiosperm. Almost all of the food eaten by humans comes from angiosperms or from animals that eat angiosperms. Grains, vegetables, herbs, and spices are just a few examples of foods that come from angiosperms. Many other items, such as clothing, medicines, and building materials, also come from these plants.

 Key Concept Check How do other organisms depend on seed plants?

Flowers

Angiosperms produce seeds that are part of a fruit. This fruit grows from parts of a flower. All angiosperms produce flowers. Some flowers, such as tulips and roses, are beautiful and showy. You also might be familiar with other flowers, such as dandelions, because you have seen them growing in your neighborhood. However, some plants produce flowers that you might never have noticed. Grass flowers are tiny and not easily seen, as shown in **Figure 19.**

 Key Concept Check How are angiosperms and gymnosperms alike, and how are they different?

Types of Angiosperms

Tulips

Grass

Beans

Math Skills

Use Percentages

A percentage compares a part to a whole. For example, out of about 1,090 species of gymnosperms, 700 species are conifers. What percentage of gymnosperms are conifers?

Express the information as a fraction.

$$\frac{700 \text{ conifers}}{1,090 \text{ gymnosperms}}$$

Change the fraction to a decimal.

$$\frac{700}{1,090} = 0.64$$

Multiply by 100 and add a percent sign.

$$0.64 \times 100 = 64\%$$

Practice

Out of 1,090 species of gymnosperms, 300 species are cycads. What percentage of gymnosperm species are cycads?

Review
- **Math Practice**
- **Personal Tutor**

Figure 19 All angiosperms have flowers that contain reproductive organs. After pollination and fertilization, seeds are produced within a flower.

Lesson 3 • **319**
EXPLAIN

Table 1 Monocots v. Dicots

Monocots	Dicots
Leaves	
narrow with parallel veins	veins are branched
Flowers	
flower parts in multiples of three	flower parts in multiples of four or five
Stems	
vascular tissue in bundles scattered throughout the stem	vascular tissue in bundles in rings
Seeds	
one cotyledon	two cotyledons

Table 1 Monocots and dicots differ in leaf, flower, stem, and seed structure.

Visual Check How are the leaves, flowers, stems, and seeds of monocots different from those of dicots?

Annuals, Biennials, and Perennials

Plants that grow, flower, and produce seeds in one growing season are called annuals. After one growing season, the plant dies. Examples include tomatoes, beans, pansies, and many common weeds.

Biennials complete their life cycles in two growing seasons. During the first year, the plant grows roots, stems, and leaves. The part of the plant that is above ground might become dormant during the winter months. In the second growing season, the plant produces new stems and leaves. It also flowers and produces seeds during this second growing season. After flowering and producing seeds, the plant dies. Carrots, beets, and foxglove are all biennials.

Perennial plants can live for more than two growing seasons. Trees and shrubs are perennials. The leaves and stems of some herbaceous perennial plants die in the winter. Stored food in the roots is used each spring for new growth.

 Reading Check How do the growing seasons of an annual, a biennial, and a perennial differ? Give an example of each type of plant.

Monocots and Dicots

Flowering plants traditionally have been organized into two groups—monocots and dicots. These groups are based on the number of leaves in early development, or cotyledons (kah tuh LEE dunz), in a seed. Researchers have learned that dicots can be organized further into two groups based on the structure of their pollen. However, because these two groups of dicots share many characteristics, we will continue to refer to them just as dicots. Look carefully at **Table 1** to learn some of the differences between monocots and dicots.

 Key Concept Check What adaptations of flowering plants enable them to survive in diverse environments?

Lesson 3 Review

Visual Summary

Angiosperms are flowering plants.

Seed plants have many adaptations that enable them to survive in diverse environments.

Seed plants have many uses.

FOLDABLES

Use your lesson Foldable to review the lesson. Save your Foldable for the project at the end of the chapter.

What do you think NOW?

You first read the statements below at the beginning of the chapter.

5. Some mosses and gymnosperms are used for commercial purposes.

6. All plants grow, flower, and produce seeds in one growing season.

Did you change your mind about whether you agree or disagree with the statements? Rewrite any false statements to make them true.

Use Vocabulary

1. The tissue that produces new xylem and phloem cells is the _____.

2. **Write** a sentence using the terms *xylem* and *phloem*.

3. **Define** *stomata* in your own words.

Understand Key Concepts

4. Which cells carry on most of a plant's photosynthesis?
   A. guard cells
   B. xylem cells
   C. palisade mesophyll cells
   D. spongy mesophyll cells

5. **Evaluate** the importance of the guard cells that surround the stomata.

6. **Contrast** gymnosperms with angiosperms.

7. **Distinguish** between a woody stem and an herbaceous stem.

Interpret Graphics

8. **Organize** Copy and fill in the table below to describe the function of each plant structure.

Structure	Function
Roots	
Stem	
Leaves	

Critical Thinking

9. **Explain** why the placement of the cells in the figure at right is so critical to the function of the phloem.

Math Skills

 Review
— Math Practice —

10. There are 300,000 species of seed plants. There are 9,000 species of grasses. What percentage of seed plants are grass species?

Inquiry Lab

Materials

computer with Internet access

Compare and Contrast Extreme Plants

Over time, plant species have developed adaptations to many environments. As you examine the types of plants that live in extreme environments, you can compare and contrast the features plants have developed that enable them to be successful in such environments.

Arctic

Ask a Question

How are plants able to live in extreme environments?

Make Observations

1. Read and complete a lab safety form.

2. Observe the environments shown in the pictures to the right. Choose one of the environments. In your Science Journal, write a description of the environment and an explanation of why you think it might be difficult for plants to grow in that environment.

Hot springs

3. Select one plant from the list of those living in the environment you chose to investigate.

Form a Hypothesis

4. After writing a description of the environment you chose, form a hypothesis explaining how a plant might be able to survive in the extreme environment.

Saltwater marsh

Test Your Hypothesis

5. Research the environment and the plant you selected. Copy the table provided. Fill out the table with as many details as you can find. As you fill in details on your plant, fill in the other half of the table with information about a common plant in your area. This will allow you to compare and contrast an extreme environment and plant with a familiar environment and plant.

6. While you research, look for photographs or illustrations of the plant you are investigating. Pay particular attention to graphics that show the unique features of your plant or illustrate how it survives in its extreme environment.

	Plant in an extreme environment	Common plant in your home environment
Name of plant		
Environment the plant lives in		
Challenges the plant face in its environment		
Resources plant needs to survive		
Plant's specialized structures to live in its environment		
Plant's method of reproduction		

Analyze and Conclude

7 **Summarize** the differences between the plant that lives in the extreme environment and the plant that lives in your familiar environment. How do these differences enable the plant living in the extreme environment to thrive in its environment?

8 **Analyze** Why is the extreme environment you studied difficult for other plants to live in? How do these conditions create an opportunity for the extreme plant?

9 **Infer** How well do you think the extreme plant would grow if you tried to grow it in your familiar environment instead of the extreme environment? Explain.

10 **THE BIG IDEA** **The Big Idea** Why are plants found in so many different environments, including extreme environments?

Communicate Your Results

Collect images of the plant and the extreme environment you researched. Create two side-by-side pages that illustrate how your plant survives in its environment. Use call-out boxes to point out the special features and structures of the plant that ensure the plant's survival.

 **Extension**

Research other plants that fill interesting niches. For example, many conifers have developed resistance to forest fires and produce seed cones that are dependent on heat from a forest fire to germinate and grow.

Lab Tips

☑ Focus on key features and structures that plants use to respond to the environment for survival when you do your research.

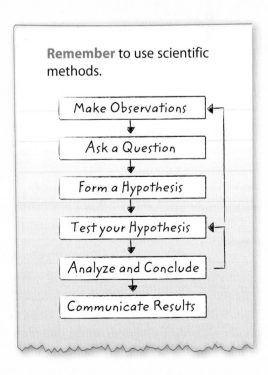

Remember to use scientific methods.

Make Observations

↓

Ask a Question

↓

Form a Hypothesis

↓

Test your Hypothesis

↓

Analyze and Conclude

↓

Communicate Results

Chapter 9 Study Guide

THE BIG IDEA

Different plant species have different adaptations that enable them to survive in most of the environments on Earth.

Key Concepts Summary 🔑

	Vocabulary
Lesson 1: What is a plant?  • Plants are multicellular **producers** composed of eukaryotic cells with cell walls composed of **cellulose.** Many plant cells have chloroplasts. • Plants have developed adaptations, such as a cell wall for support, a **cuticle** to prevent water loss and to provide protection from insects, **vascular tissue** to transport materials, and numerous reproductive strategies, to survive in Earth's changing environments. • Members of the plant kingdom are classified into groups called divisions, which are equivalent to phyla in other kingdoms. Plants have two-word scientific names.	**producer** p. 298 **cuticle** p. 299 **cellulose** p. 299 **vascular tissue** p. 300
Lesson 2: Seedless Plants • Vascular and nonvascular seedless plants are multicellular producers composed of eukaryotic cells. Nonvascular seedless plants usually are smaller than vascular seedless plants, lack vascular tissue, and have **rhizoids** instead of roots to anchor them.	**rhizoid** p. 307 **frond** p. 309
Lesson 3: Seed Plants • All seed plants make seeds and reproduce. Seed plants have leaves, stems, roots, and vascular tissue—**xylem** and **phloem.** • Seed plants are important to other organisms for various reasons including for food, for the addition of oxygen to the environment, and for commercial uses. • Gymnosperms and angiosperms are both seed plants. Angiosperms produce flowers, gymnosperms do not. The seeds of angiosperms are surrounded by fruit. The seeds of gymnosperms are not surrounded by fruit. • Flowering plants have adaptations that enable them to survive in diverse environments. Such adaptations include leaves, stems, vascular tissue, roots, flowers, and seeds protected by a fruit.	**cambium** p. 314 **xylem** p. 314 **phloem** p. 315 **stoma** p. 317

FOLDABLES® Chapter Project

Assemble your lesson Foldables as shown to make a Chapter Project. Use the project to review what you have learned in this chapter.

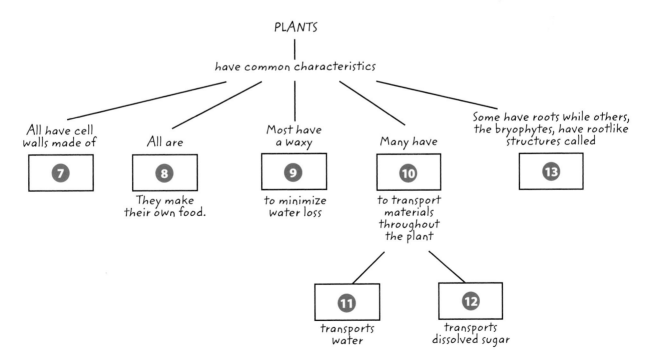

Common Characteristics of Plants

Adaptations to the Environment

Vascular Seedless Plants

Both

Nonvascular Seedless Plants

Gymnosperms

Both

Angiosperms

Use Vocabulary

1. Distinguish between xylem and phloem.

2. The openings in leaves that allow gases to pass into and out of the leaf are _____.

3. Define the term *rhizoid* in your own words.

4. The tissue that produces new xylem and phloem cells is the _____.

5. Write a sentence using the term *cuticle*.

6. Use the term *vascular tissue* in a sentence.

Link Vocabulary and Key Concepts

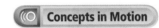 **Concepts in Motion** Interactive Concept Map

Copy this concept map, and then use vocabulary terms from the previous page to complete the concept map.

PLANTS

have common characteristics

All have cell walls made of

7

All are

8

They make their own food.

Most have a waxy

9

to minimize water loss

Many have

10

to transport materials throughout the plant

Some have roots while others, the bryophytes, have rootlike structures called

13

11

transports water

12

transports dissolved sugar

Chapter 9 Review

Understand Key Concepts

1 Guard cells control the size of the opening of the
A. cambium.
B. cotyledon.
C. stomata.
D. xylem.

2 The major function of leaves is to
A. anchor the plant.
B. perform photosynthesis.
C. shade the tree.
D. support the stem.

3 Which is an angiosperm?
A. fern
B. moss
C. pine
D. tulip

4 Which does the plant below NOT have?

A. chloroplasts
B. rhizoids
C. cell walls
D. vascular tissue

5 Which is the plant part responsible for anchoring the plant in soil?
A. flower
B. leaf
C. root
D. stem

6 Which are vascular plants?
A. ferns
B. hornworts
C. liverworts
D. mosses

7 Only members of the plant kingdom are organized into
A. categories.
B. divisions.
C. groups.
D. phyla.

8 Which is composed of cellulose?
A. chloroplasts
B. cytoplasm
C. plant cell membrane
D. plant cell wall

9 What is the function of the stomata?
A. to perform photosynthesis
B. to produce sugar
C. to allow water into the leaf
D. to enable gases to enter and leave

10 Which are produced by angiosperms but not by gymnosperms?
A. cones
B. flowers
C. leaves
D. seeds

11 The leaflike structures of mosses do NOT contain
A. chloroplasts.
B. water.
C. photosynthetic cells.
D. vascular tissue.

12 What is the function of the structure shown below?

A. control gas exchange
B. perform photosynthesis
C. transport sugar
D. transport water

Critical Thinking

13 **Choose** one of the adaptations that plants have for living on land, and explain its significance.

14 **Suggest** an environment where plants would not need a cuticle.

15 **Design** a new division of plants. Describe them and their environment.

16 **Suggest** additional uses for peat moss.

17 **Construct** a table to organize information about vascular seedless plants.

18 **Evaluate** the lack of cuticle in moss plants.

19 **Explain** the advantage of fruit production in angiosperms.

20 **Predict** the impact of a disease that killed all gymnosperms. Explain your reasoning.

21 **Explain** how the structure of a leaf as shown in the figure below is appropriate for its role in photosynthesis.

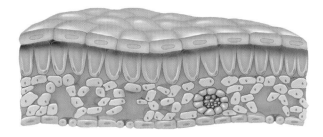

Writing in Science

22 **Write** a short story about a hiker's thoughts as he or she comes across different plants such as mosses, liverworts, ferns, horsetails, gymnosperms, and angiosperms.

REVIEW THE BIG IDEA

23 Write a brief description of two different environments. Now describe how plant adaptations would help the plants survive in each of your environments.

24 What structures enable the plants in the photo below to live in this environment?

Math Skills

Review
— Math Practice —

Use Percentages

25 Some scientists estimate the total number of species of organisms on Earth to be about 14,000,000. If there are 300,000 species of seed plants, what percentage of Earth's species are seed plants?

26 Of the 1,090 species of gymnosperms, there is only one species of ginkgo. What percentage of gymnosperm species are ginkgoes?

27 Of the 1,090 species of gymnosperms, 90 species are gnetophytes. What percentage of gymnosperm species are gnetophytes?

Standardized Test Practice

Record your answers on the answer sheet provided by your teacher or on a sheet of paper.

Multiple Choice

1 Which is NOT a characteristic of all plants?

   A They are multicellular.

   B They have vascular tissue.

   C They make their own food.

   D They undergo photosynthesis.

2 Which are seedless nonvascular plants?

   A conifers

   B grasses

   C mosses

   D tulips

Use the diagram below to answer question 3.

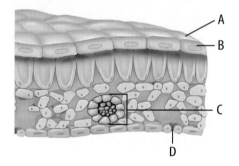

3 Which feature in the figure is a plant adaptation to life on land and reduces water loss through evaporation?

   A A

   B B

   C C

   D D

4 Which are NOT common to all seed plants?

   A cell walls

   B flowers

   C roots

   D vascular tissues

5 Which feature do scientists use to classify angiosperms as monocots or dicots?

   A embryonic leaves

   B flower shapes

   C seed numbers

   D xylem cells

Use the image below to answer question 6.

6 Which plants form seeds in structures such as the one shown in the figure?

   A biennials

   B conifers

   C maples

   D roses

7 Which plants provide most of the food humans eat?

   A angiosperms

   B gymnosperms

   C seedless nonvascular plants

   D seedless vascular plants

8 Which structures are in plant cells but are not in animal cells?

   A cell membranes and vacuoles

   B cell walls and chloroplasts

   C chloroplasts and mitochondria

   D vacuoles and mitochondria

Use the figure below to answer question 9.

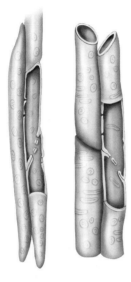

9 What type of plant tissue contains the cells shown in the figure?

   A cambium

   B mesophyll

   C phloem

   D xylem

10 An unknown plant is a few centimeters tall when fully grown. It lives only in moist areas and reproduces with spores. How would you classify this plant?

   A nonvascular seed plant

   B nonvascular seedless plant

   C vascular seedless plant

   D vascular seed plant

Constructed Response

Use the figure below to answer questions 11 and 12.

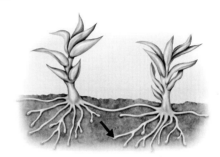

11 The above diagram shows a seedless nonvascular plant. Identify the structure marked with an arrow, and explain its function.

12 Which structure in vascular plants performs a similar function?

13 Some angiosperms can store food in their roots. How does this adaptation help them to survive diverse environmental conditions? Give an example of a plant that has this adaptation.

14 How do the leaves of gymnosperms differ from the leaves of angiosperms?

15 Describe the role of vascular tissue in enabling a plant to adapt to its environment.

NEED EXTRA HELP?															
If You Missed Question...	1	2	3	4	5	6	7	8	9	10	11	12	13	14	15
Go to Lesson...	1	2	1	3	1	3	3	1	3	1 & 2	2	2 & 3	3	3	1

Plant Processes and Reproduction

THE BIG IDEA What processes enable plants to survive and reproduce?

Inquiry Holding on for Dear Life?

The tendril of this *Omphalea* (om FAL ee uh) vine grows around a branch in a tropical rain forest.

- How do you think growing around another plant might help the *Vicia* plant survive?

- Can you think of any other processes that enable plants to survive and reproduce?

Get Ready to Read

What do you think?

Before you read, decide if you agree or disagree with each of these statements. As you read this chapter, see if you change your mind about any of the statements.

1. Plants do not carry on cellular respiration.

2. Plants are the only organisms that carry on photosynthesis.

3. Plants do not produce hormones.

4. Plants can respond to their environments.

5. Seeds contain tiny plant embryos.

6. Flowers are needed for plant reproduction.

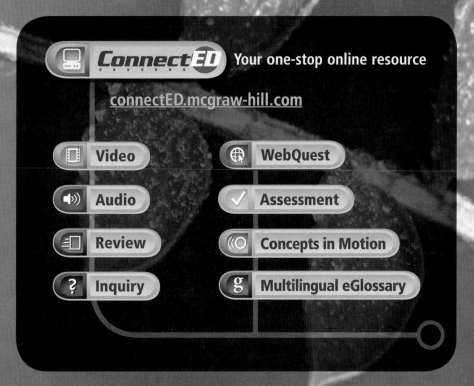

ConnectED Your one-stop online resource

connectED.mcgraw-hill.com

- Video
- Audio
- Review
- Inquiry
- WebQuest
- Assessment
- Concepts in Motion
- Multilingual eGlossary

Lesson 1

Reading Guide

Key Concepts 🔑
ESSENTIAL QUESTIONS

- How do materials move through plants?
- How do plants perform photosynthesis?
- What is cellular respiration?
- What is the relationship between photosynthesis and cellular respiration?

Vocabulary
photosynthesis p. 334
cellular respiration p. 336

g Multilingual eGlossary

Energy Processing in Plants

Inquiry All Leaf Cells?

You are looking at a magnified cross section of a leaf. As you can see, the cells in the middle of the leaf are different from the cells on the edges. What do you think this might have to do with the cellular processes a leaf carries out that enable a plant's survival?

How can you show the movement of materials inside a plant?

Most parts of plants need water. They also need a system to move water throughout the plant so cells can use it for plant processes.

1. Read and complete a lab safety form.

2. Gently pull two stalks from the base of a bunch of **celery.** Leave one stalk complete. Use a **paring knife** to carefully cut directly across the bottom of the second stalk.

3. Put 100 mL of water into each of two **beakers.** Place 3–4 drops of **blue food coloring** into the water. Place one celery stalk in each beaker.

4. After 20 min, observe the celery near the bottom of each stalk. Observe again after 24 h. Record your observations in your Science Journal.

Think About This

1. What happened in each celery stalk?

2. **Key Concept** What did the colored water do? Why do you think this occurred?

Materials for Plant Processes

Food, water, and oxygen are three things you need to survive. Some of your organ systems process these materials, and others transport them throughout your body. Like you, plants need food, water, and oxygen to survive. Unlike you, plants do not take in food. Most of them make their own.

Moving Materials Inside Plants

You might recall reading about xylem (ZI lum) and phloem (FLOH em)—the vascular tissue in most plants. These tissues transport materials throughout a plant.

After water enters a plant's roots, it moves into xylem. Water then flows inside xylem to all parts of a plant. Without enough water, plant cells wilt, as shown in **Figure 1.**

Most plants make their own food—a liquid sugar. The liquid sugar moves out of food-making cells, enters phloem, and flows to all plant cells. Cells break down the sugar and release energy. Some plant cells can store food.

Plants require oxygen and carbon dioxide to make food. Like you, plants produce water vapor as a waste product. Carbon dioxide, oxygen, and water vapor pass into and out of a plant through tiny openings in leaves.

 **Key Concept Check** How do materials move through plants?

Figure 1 This plant wilted due to lack of water in the soil.

WORD ORIGIN ············

photosynthesis
from Greek *photo-*, means "light"; and *synthesis*, means "composition"

Figure 2 Photosynthesis occurs inside the chloroplasts of mesophyll cells in most leaves.

Photosynthesis

Plants need food, but they cannot eat as people do. They make their own food, and leaves are the major food-producing organs of plants. This means that leaves are the sites of photosynthesis (foh toh SIHN thuh sus). **Photosynthesis** *is a series of chemical reactions that convert light energy, water, and carbon dioxide into the food-energy molecule glucose and give off oxygen.* The structure of a leaf is well-suited to its role in photosynthesis.

Leaves and Photosynthesis

As shown in **Figure 2,** leaves have many types of cells. The cells that make up the top and bottom layers of a leaf are flat, irregularly shaped cells called epidermal (eh puh DUR mul) cells. On the bottom epidermal layer of most leaves are small openings called stomata (STOH muh tuh). Carbon dioxide, water vapor, and oxygen pass through stomata. Epidermal cells can produce a waxy covering called the cuticle.

Most photosynthesis occurs in two types of mesophyll (ME zuh fil) cells inside a leaf. These cells contain chloroplasts, the organelle where photosynthesis occurs. Near the top surface of the leaf are palisade mesophyll cells. They are packed together. This arrangement exposes the most cells to light. Spongy mesophyll cells have open spaces between them. Gases needed for photosynthesis flow through the spaces between the cells.

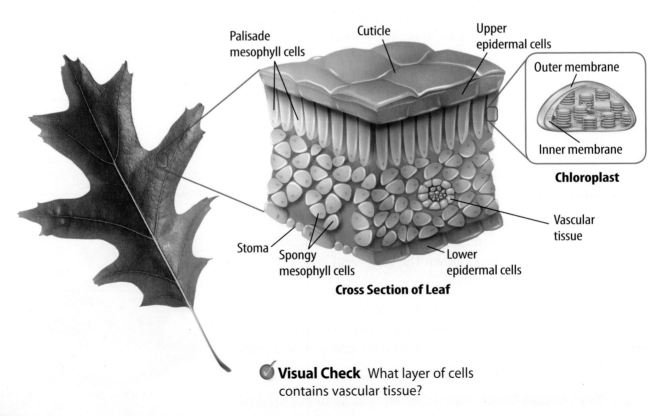

Palisade mesophyll cells · Cuticle · Upper epidermal cells · Outer membrane · Inner membrane · **Chloroplast** · Vascular tissue · Stoma · Spongy mesophyll cells · Lower epidermal cells

Cross Section of Leaf

✓ **Visual Check** What layer of cells contains vascular tissue?

Capturing Light Energy

As you read about the steps of photosynthesis, refer to **Figure 3** to help you understand the process. In the first step of photosynthesis, plants capture the energy in light. This occurs in chloroplasts. Chloroplasts contain plant pigments. Pigments are chemicals that can absorb and reflect light. Chlorophyll, the most common plant pigment, is necessary for photosynthesis. Most plants appear green because chlorophyll reflects green light. Chlorophyll absorbs other colors of light. This light energy is used during photosynthesis.

Once chlorophyll traps and stores light energy, this energy can be transferred to other molecules. During photosynthesis, water molecules are split apart. This releases oxygen into the atmosphere, as shown in **Figure 3**.

 Reading Check How do plants capture light energy?

Making Sugars

Sugars are made in the second step of photosynthesis. This step can occur without light. In chloroplasts, carbon dioxide from the air is converted into sugars by using the energy stored and trapped by chlorophyll. Carbon dioxide combines with hydrogen atoms from the splitting of water molecules and forms sugar molecules. Plants can use this sugar as an energy source or can store it. Potatoes and carrots are examples of plant structures where excess sugar is stored.

Key Concept Check What are the two steps of photosynthesis?

Why is photosynthesis important?

Try to imagine a world without plants. How would humans or other animals get the oxygen that they need? Plants help maintain the atmosphere you breathe. Photosynthesis produces most of the oxygen in the atmosphere.

Photosynthesis 🔑

Figure 3 Photosynthesis is a series of complex chemical processes. The first step is capturing light energy. In the second step, that energy is used for making glucose, a type of sugar.

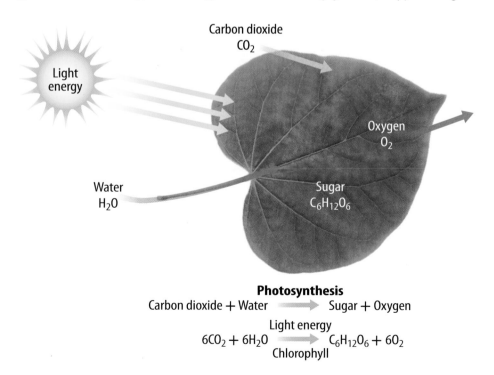

Carbon dioxide CO_2

Light energy

Water H_2O

Oxygen O_2

Sugar $C_6H_{12}O_6$

Photosynthesis

Carbon dioxide + Water ⟶ Sugar + Oxygen

Light energy

$6CO_2 + 6H_2O$ ⟶ $C_6H_{12}O_6 + 6O_2$

Chlorophyll

Cellular Respiration

All organisms require **energy** to survive. Energy is in the chemical bonds in food molecules. A process called cellular respiration releases energy. **Cellular respiration** *is a series of chemical reactions that convert the energy in food molecules into a usable form of energy called ATP.*

Releasing Energy from Sugars

Glucose molecules break down during cellular respiration. Much of the energy released during this process is used to make ATP, an energy storage molecule. This process requires oxygen, produces water and carbon dioxide as waste products, and occurs in the cytoplasm and mitochondria of cells.

Why is cellular respiration important?

If your body did not break down the food you eat through cellular respiration, you would not have energy to do anything. Plants produce sugar, but without cellular respiration, plants could not grow, reproduce, or repair tissues.

 Key Concept Check What is cellular respiration?

Inquiry MiniLab   **20 minutes**

Can you observe plant processes?

Plants perform both photosynthesis and cellular respiration. Can you observe both processes in radish seedlings?

1. Read and complete a lab safety form.
2. Put **potting soil** in the bottom of a **small, self-sealing plastic bag** so that the soil is 3–4 cm deep. Dampen the soil.
3. Drop several **radish seeds** into the bag and close the top, but leave a small opening so air can still get into the bag.
4. Place the bag upright in a place that has a **light source.** Each group should use a different light source. Observe for 4–5 days.
5. Carefully place an open container of **bromthymol blue (0.004%) solution** upright in the bag next to the seedlings. Bromthymol blue turns yellow in the presence of carbon dioxide.
6. Seal the bag. Observe the bag and its contents the next day. Record your observations in your Science Journal.

Analyze and Conclude

1. **Describe** the differences in seedling samples among groups. Why are there differences?
2. **Evaluate** What change in the bromthymol blue solution did you observe? Why?
3. **Key Concept** Explain what processes occurred in the seedlings.

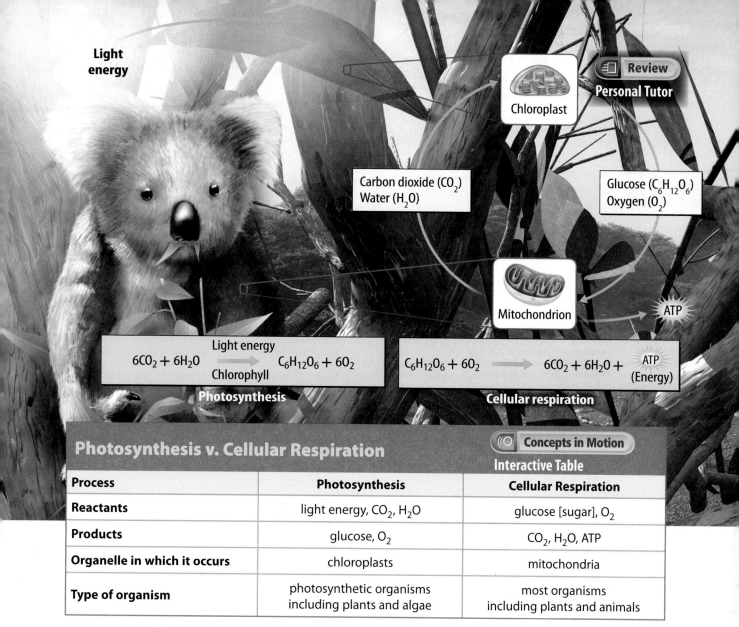

Light energy

Chloroplast

Carbon dioxide (CO_2)
Water (H_2O)

Glucose ($C_6H_{12}O_6$)
Oxygen (O_2)

Review
Personal Tutor

Mitochondrion

ATP

$$6CO_2 + 6H_2O \xrightarrow[\text{Chlorophyll}]{\text{Light energy}} C_6H_{12}O_6 + 6O_2$$

Photosynthesis

$$C_6H_{12}O_6 + 6O_2 \longrightarrow 6CO_2 + 6H_2O + \begin{array}{c}\text{ATP}\\(\text{Energy})\end{array}$$

Cellular respiration

Photosynthesis v. Cellular Respiration

Concepts in Motion
Interactive Table

Process	Photosynthesis	Cellular Respiration
Reactants	light energy, CO_2, H_2O	glucose [sugar], O_2
Products	glucose, O_2	CO_2, H_2O, ATP
Organelle in which it occurs	chloroplasts	mitochondria
Type of organism	photosynthetic organisms including plants and algae	most organisms including plants and animals

Comparing Photosynthesis and Cellular Respiration

Photosynthesis requires light energy and the reactants—substances that react with one another during the process—carbon dioxide and water. Oxygen and the energy-rich molecule glucose are the products, or end substances, of photosynthesis. Most plants, some protists, and some bacteria carry on photosynthesis.

Cellular respiration requires the reactants glucose and oxygen, produces carbon dioxide and water, and releases energy in the form of ATP. Most organisms carry on cellular respiration. Photosynthesis and cellular respiration are interrelated, as shown in **Figure 4.** Life on Earth depends on a balance of these two processes.

Key Concept Check How are photosynthesis and cellular respiration alike, and how are they different?

Figure 4 The relationship between cellular respiration and photosynthesis is important for life.

Visual Check What are the reactants of cellular respiration? What are the products?

Lesson 1 Review

Visual Summary

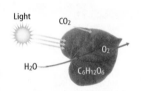

Materials that a plant requires to survive move through the plant in the vascular tissue, xylem and phloem.

Plants can make their own food by using light energy, water, and carbon dioxide.

The products of photosynthesis are the reactants for cellular respiration.

 FOLDABLES®

Use your lesson Foldable to review the lesson. Save your Foldable for the project at the end of the chapter.

What do you think NOW?

You first read the statements below at the beginning of the chapter.

1. Plants do not carry on cellular respiration.

2. Plants are the only organisms that carry on photosynthesis.

Did you change your mind about whether you agree or disagree with the statements? Rewrite any false statements to make them true.

Use Vocabulary

1. A series of chemical reactions that convert the energy in food molecules into a usable form of energy, called ATP, is called _____.

2. **Define** *photosynthesis* in your own words.

Understand Key Concepts 🔑

3. Which structure moves water through plants?
   - **A.** chloroplast
   - **C.** nucleus
   - **B.** mitochondrion
   - **D.** xylem

4. **Describe** how plants use chlorophyll for photosynthesis.

5. **Summarize** the process of cellular respiration.

Interpret Graphics

6. **Explain** how the structure shown below is organized for its role in photosynthesis.

7. **Compare and Contrast** Copy and fill in the table below to compare and contrast photosynthesis and cellular respiration.

Process	Similarities	Differences

Critical Thinking

8. **Predict** the effect of a plant disease that destroys all of the chloroplasts in a plant.

9. **Evaluate** why plants perform cellular respiration.

Deforestation and Carbon Dioxide
in the Atmosphere

How does carbon dioxide affect climate?

What do you think when you hear the words *greenhouse gases*? Many people picture pollution from automobiles or factory smokestacks. It might be surprising to learn that cutting down forests affects the amount of one of the greenhouse gases in the atmosphere—carbon dioxide.

Deforestation is the term used to describe the destruction of forests. Deforestation happens because people cut down forests to use the land for other purposes, such as agriculture or building sites, or to use the trees for fuel or building materials.

Trees, like most plants, carry out photosynthesis and make their own food. Carbon dioxide from the atmosphere is one of the raw materials, or reactants, of photosynthesis. When deforestation occurs, trees are unable to remove carbon dioxide from the atmosphere. As a result, the level of carbon dioxide in the atmosphere increases.

Trees affect the amount of atmospheric carbon dioxide in other ways. Large amounts of carbon are stored in the molecules that make up trees. When trees are burned or left to rot, much of this stored carbon is released as carbon dioxide. This increases the amount of carbon dioxide in the atmosphere.

Carbon dioxide in the atmosphere has an impact on climate. Greenhouse gases, such as carbon dioxide, increase the amount of the Sun's energy that is absorbed by the atmosphere. They also reduce the ability of heat to escape back into space. So, when levels of carbon dioxide in the atmosphere increase, more heat is trapped in Earth's atmosphere. This can lead to climate change.

▲ These cattle are grazing on land that was once part of a forest in Brazil.

▲ In a process called slash-and-burn, forest trees are cut down and burned to clear land for agriculture.

It's Your Turn

RESEARCH AND REPORT How can we lower the rate of deforestation? What are some actions you can take that could help slow the rate of deforestation? Research to find out how you can make a difference. Make a poster to share what you learn.

Reading Guide

Key Concepts

ESSENTIAL QUESTIONS

- How do plants respond to environmental stimuli?
- How do plants respond to chemical stimuli?

Vocabulary

stimulus p. 341

tropism p. 342

photoperiodism p. 344

plant hormone p. 345

 Multilingual eGlossary

Video **BrainPOP®**

Plant Responses

Inquiry A Meat-Eating Plant?

Venus flytraps have leaves that look like jaws. The leaves close only when a stimulus, such as a fly, brushes against tiny, sensitive hairs on the surface of the leaves. To what other stimuli do you think plants might respond?

How do plants respond to stimuli?

Plants use light energy and make their own food during photosynthesis. How else do plants respond to light in their environment?

1 Read and complete a lab safety form.

2 Choose a **pot of young radish seedlings.**

3 Place **toothpicks** parallel to a few of the seedlings in the pot in the direction of growth.

4 Place the pot near a **light source**, such as a gooseneck lamp or next to a window. The light source should be to one side of the pot, not directly above the plants.

5 Check the position of the seedlings in relation to the toothpicks after 30 minutes. Record your observations in your Science Journal.

6 Observe the seedlings when you come to class the next day. Record your observations.

Think About This

1. What happened to the position of the seedlings after the first 30 minutes? What is your evidence of change?

2. What happened to the position of the seedlings after a day?

3. 🔑 **Key Concept** Why do you think the position of the seedlings changed?

Stimuli and Plant Responses

Have you ever been in a dark room when someone suddenly turned on the light? You might have reacted by quickly shutting or covering your eyes. **Stimuli** (STIM yuh li; singular, stimulus) *are any changes in an organism's environment that cause a response.*

Often a plant's response to stimuli might be so slow that it is hard to see it happen. The response might occur gradually over a period of hours or days. Light is a stimulus. A plant responds to light by growing toward it, as shown in **Figure 5**. This response occurs over several hours.

In some cases, the response to a stimulus is quick, such as the Venus flytrap's response to touch. When stimulated by an insect's touch, the two sides of the trap snap shut immediately, trapping the insect inside.

✓ **Reading Check** Why is it sometimes hard to see a plant's response to a stimulus?

Figure 5 The light is the stimulus, and the seedlings have responded by growing toward the light.

Environmental Stimuli

When it is cold outside, you probably wear a sweatshirt or a coat. Plants cannot put on warm clothes, but they do respond to their environments in a variety of ways. You might have seen trees flower in the spring or drop their leaves in the fall. Both are plant responses to environmental stimuli.

Growth Responses

Plants respond to a number of different environmental stimuli. These include light, touch, and gravity. *A* **tropism** *(TROH pih zum) is a response that results in plant growth toward or away from a stimulus.* When the growth is toward a stimulus, the tropism is called positive. A plant bending toward light is a positive tropism. Growth away from a stimulus is considered negative. A plant's stem growing upward against gravity is a negative tropism.

Light The growth of a plant toward or away from light is a tropism called phototropism. A plant has a light-sensing chemical that helps it detect light. Leaves and stems tend to grow in the direction of light, as shown in **Figure 6.** This response maximizes the amount of light the plant's leaves receive. Roots generally grow away from light. This usually means that the roots grow down into the soil and help anchor the plant.

 Reading Check How is phototropism beneficial to a plant?

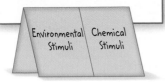

WORD ORIGIN · · · · · · · · · · ·

tropism
from Greek *tropos,* means "turn" or "turning"

· · · · · · · · · · · · · · · · ·

Response to Light 🔑

Figure 6 As a plant's leaves turn toward the light, the amount of light that the leaves can absorb increases.

Touch The response of a plant to touch is called a thigmotropism (thihg MAH truh pih zum). You might have seen vines growing up the side of a building or a fence. This happens because the plant has special structures that respond to touch. These structures, called tendrils, can wrap around or cling to objects, as shown in **Figure 7.** A tendril wrapping around an object is an example of positive thigmotropism. Roots display negative thigmotropism. They grow away from objects in soil, enabling them to follow the easiest path through the soil.

Gravity The response of a plant to gravity is called gravitropism. Stems grow away from gravity, while roots grow toward gravity. The seedlings in **Figure 8** are exhibiting both responses. No matter how a seed lands on soil, when it starts to grow, its roots grow down into the soil. The stem grows up. This happens even when a seed is grown in a dark chamber, indicating that these responses can occur independently of light.

 Key Concept Check What types of environmental stimuli do plants respond to? Give three examples.

Response to Touch 🔑

▲ **Figure 7** The tendrils of the vine respond to touch and coil around the branch.

Response to Gravity 🔑

Figure 8 Both of these plant stems are growing away from gravity. The upward growth of a plant's stem is negative gravitropism, and the downward growth of its roots is positive gravitropism.

🔵 **Visual Check** How is the plant on the left responding to the pot begin placed on its side?

Flowering Responses

You might think all plants respond to light, but in some plants, flowering is actually a response to darkness! **Photoperiodism** *is a plant's response to the number of hours of darkness in its environment.* Scientists once hypothesized that photoperiodism was a response to light. Therefore, these flowering responses are called long-day, short-day, and day-neutral and relate to the number of hours of daylight in a plant's environment.

Long-Day Plants Plants that flower when exposed to less than 10–12 hours of darkness are called long-day plants. The carnations shown in **Figure 9** are examples of long-day plants. This plant usually produces flowers in summer, when the number of hours of daylight is greater than the number of hours of darkness.

Short-Day Plants Short-day plants require 12 or more hours of darkness for flowering to begin. An example of a short-day plant is the poinsettia, shown in **Figure 9.** Poinsettias tend to flower in late summer or early fall when the number of hours of daylight is decreasing and the number of hours of darkness is increasing.

Day-Neutral Plants The flowering of some plants doesn't seem to be affected by the number of hours of darkness. Day-neutral plants flower when they reach maturity and the environmental conditions are right. Plants such as the roses in **Figure 9** are day-neutral plants.

 Reading Check How is the flowering of day-neutral plants affected by exposure to hours of darkness?

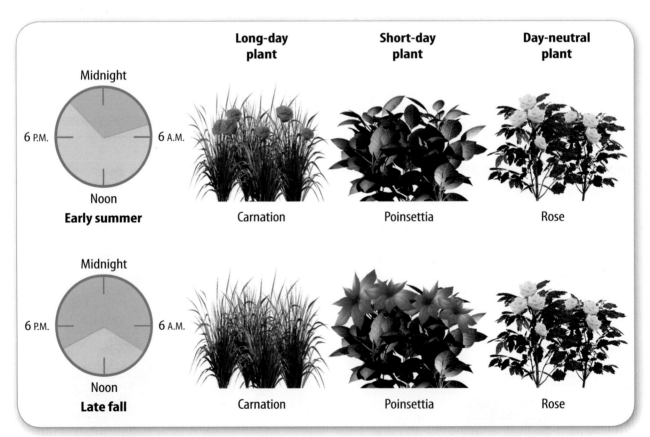

Figure 9 The number of hours of darkness controls flowering in many plants. Long-day plants flower when there are more hours of daylight than darkness, and short-day plants flower when there are more hours of darkness than daylight.

Visual Check What time of year receives more darkness, and what type of plant produces flowers during that season?

Chemical Stimuli

Plants respond to chemical stimuli as well as environmental stimuli. **Plant hormones** *are substances that act as chemical messengers within plants.* These chemicals are produced in tiny amounts. They are called messengers because they usually are produced in one part of a plant and affect another part of that plant.

Auxins

One of the first plant hormones discovered was auxin (AWK sun). There are many different kinds of auxins. Auxins generally cause increased plant growth. They are responsible for phototropism, the growth of a plant toward light. Auxins concentrate on the dark side of a plant's stem, and these cells grow longer. This causes the stem of the plant to grow toward the light, as shown in **Figure 10.**

Ethylene

The plant hormone ethylene helps stimulate the ripening of fruit. Ethylene is a gas that can be produced by fruits, seeds, flowers, and leaves. You might have heard someone say that one rotten apple spoils the whole barrel. This is based on the fact that rotting fruits release ethylene. This can cause other fruits nearby to ripen and possibly rot. Ethylene also can cause plants to drop their leaves.

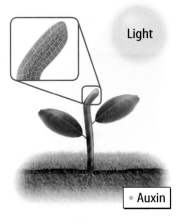

Figure 10 Auxin on the left side of the seedling causes more growth and makes the seedling bend to the right.

Key Concept Check How do plants respond to the chemical stimuli, or hormones, auxin and ethylene?

 MiniLab

20 minutes

When will plants flower?

Did you ever think plants could have strategies so that they can germinate, live, grow, reproduce, and continue their species? Photoperiodism is one such strategy.

1. In your Science Journal, copy the table below to classify plants based on their photoperiodisms.

2. Choose 8–10 **pictures of flowers.** Record their names in your table. Use the clues on the back of each photo to determine the correct photoperiodism of each plant.

Analyze and Conclude

1. **Interpret Data** Based on your table, which plants would flower during the summer?

2. **Explain** why some plants flower at the same time every year.

3. **Infer** what might happen if short-day plants were placed under light for an hour or two at night.

4. **Key Concept** Why would photoperiodism be an important strategy for flowering plants?

Plant	Season	Short-Day	Long-Day	Day-Neutral

Response to Gibberellins

Gibberellins and Cytokinins

Rapidly growing areas of a plant, such as roots and stems, produce gibberellins (jih buh REL unz). These hormones increase the rate of cell division and cell elongation. This results in increased growth of stems and leaves. Gibberellins also can be applied to the outside of plants. As shown in **Figure 11,** applying gibberellins to plants can have a dramatic effect.

Root tips produce most of the cytokinins (si tuh KI nunz), another type of hormone. Xylem carries cytokinins to other parts of a plant. Cytokinins increase the rate of cell division, and in some plants, cytokinins slow the aging process of flowers and fruits.

Summary of Plant Hormones

Plants produce many different hormones. The hormones you have just read about are groups of similar compounds. Often, two or more hormones interact and produce a plant response. Scientists are still discovering new information about plant hormones.

Humans and Plant Responses

Humans depend on plants for food, fuel, shelter, and clothing. Humans make plants more productive using plant hormones. Some crops now are easier to grow because humans understand how they respond to hormones. As you study **Figure 12** on the next page, make a list of all the ways humans can benefit from understanding and using plant responses.

Reading Check How are humans dependent on plants?

Figure 12 Understanding how plants respond to hormones can benefit people in many ways.

The cutting on the left has been treated with synthetic auxins, which encourage cuttings to root.

By choosing seeds that produce climbing cucumbers, farmers grow plants that are easier to pick. The cucumbers grow faster and bigger because they get more light.

Removing the apical bud of a plant suppresses auxin causing the plant to grow out instead of up, producing a fuller plant.

Bananas can be picked and shipped while still green and then be treated with ethylene to cause them to ripen.

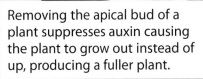

The use of cytokinins helps scientists and horticulturists grow hundreds of identical plants.

Visual Summary

Plants respond to stimuli in their environments in many ways.

Carnation

Photoperiodism occurs in long-day plants and short-day plants. Day-neutral plants are not affected by the number of hours of darkness.

Light

Auxin

Plant hormones are internal chemical stimuli that produce different responses in plants.

FOLDABLES

Use your lesson Foldable to review the lesson. Save your Foldable for the project at the end of the chapter.

What do you think NOW?

You first read the statements below at the beginning of the chapter.

3. Plants do not produce hormones.

4. Plants can respond to their environments.

Did you change your mind about whether you agree or disagree with the statements? Rewrite any false statements to make them true.

Use Vocabulary

1 **Define** *plant hormone* in your own words.

2 The response of an organism to the number of hours of darkness in its environment is called _____.

3 **Distinguish** between *stimuli* and *tropism*.

Understand Key Concepts

4 **Describe** an example of a plant responding to environmental stimuli.

5 **Distinguish** between a long-day plant and a short-day plant.

6 **Compare** the effect of auxins and gibberellins on plant cells.

7 Which is NOT likely to cause a plant response?

    A. changing the amount of daylight
    B. moving plants away from each other
    C. treating with plant hormones
    D. turning a plant on its side

Interpret Graphics

8 **Identify** Copy the table below and list the plant hormones mentioned in this lesson. Describe the effect of each on plants.

Hormone	Effect on Plants

Critical Thinking

9 **Infer** why the plant shown to the right is growing at an angle.

Math Skills

 Review
— Math Practice —

10 When sprayed with gibberellins, the diameter of mature grapes increased from 1.0 cm to 1.75 cm. What was the percent increase in size?

What happens to seeds if you change the intensity of light?

Materials

plastic tub

potting soil

fast-growing grass seeds

sun shields

light source

metric ruler

mister bottle with water

Safety

Seeds require light, water, gases, and soil to germinate, grow into seedlings, and then grow into mature plants. Different types of seeds require different amounts of each of these factors. What happens if one of these factors is changed?

Learn It

In any experiment, it is important to keep everything the same except for the item you are testing. The one factor you change, or **manipulate,** is called the independent **variable.** Your experiment should also have a control. The control is an individual instance or experimental subject for which the independent **variable** is not changed.

Try It

1 Read and complete a lab safety form.

2 Fill the plastic tub with potting soil. Water the soil, and then add more soil. Level it to about 1–2 cm from the top. Spread the grass seeds evenly across the soil. Cover the seeds with a thin layer of soil.

3 Obtain the precut shields of vellum, plastic needlepoint grid, and cardboard. These will be used to change the intensity of light coming onto the soil.

4 Cover the soil with the shields by laying them next to each other. Leave one section of soil uncovered.

5 Place the tub on a windowsill or under a growing light.

6 Keep the soil damp, not wet, with a mister. Water gently so the seeds stay in position.

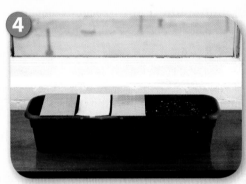

7 Design a table to record observations in your Science Journal. Include columns for day, growth pattern, height, and random sampling counts. Begin observations when seedlings first emerge. Observe seedlings for 3–5 days.

Apply It

8 **Identify** the variables and the controls used in this investigation.

9 **Analyze** the data you collected through your observations. Which light intensity appeared to bring about the fullest, tallest growth?

10 **Draw conclusions** about what would happen if you put one section of seeds in total darkness. Would it germinate? If you changed the light intensity immediately after the seeds germinated, would it survive?

11 🔑 **Key Concept** Does the amount of light affect the germination and growth of grass seeds? Explain.

Plant Reproduction

Reading Guide

Key Concepts 🔑

ESSENTIAL QUESTIONS

- What is the alternation of generations in plants?
- How do seedless plants reproduce?
- How do seed plants reproduce?

Vocabulary

alternation of generations p. 352

spore p. 352

pollen grain p. 354

pollination p. 354

ovule p. 354

embryo p. 354

seed p. 354

stamen p. 356

pistil p. 356

ovary p. 356

fruit p. 357

g Multilingual eGlossary

Inquiry A Bee's-Eye View?

Bees can see ultraviolet (UV) light. We see a dandelion as yellow. Because of a bee's ability to see UV light, a bee sees a dandelion like the one above. Why do you think bees see flowers differently than we do? Why do some plants produce flowers while others do not?

How can you identify fruits?

Flowering plants grow from seeds that they produce. Animals depend on flowering plants for food. The function of the fruit is to disperse the seeds for plant reproduction.

1. Read and complete a lab safety form.

2. Make a two-column table in your Science Journal. Label the columns *Fruits* and *Not Fruits*.

3. Examine a collection of **food items.** Determine whether each item is a fruit. Record your observations in your table.

4. Place each food item on a piece of **plastic wrap.** Use a **plastic or paring knife** to cut the items in half.

5. Examine the inside of each food item. Record your observations.

Think About This

1. What observations did you make about the insides of the food items? Would you reclassify any food item based on your observations? Explain.

2. How can the number of seeds or how they are placed in the fruit help with seed dispersal?

3. 🔑 **Key Concept** What role do you think a fruit has in a flowering plant's reproduction?

Asexual Reproduction Versus Sexual Reproduction

In early spring, you might see cars or sidewalks covered with a thick, yellow dust. Where did it come from? It probably came from plants that are reproducing. As in all living things, reproduction is part of the life cycles of plants.

Plants can reproduce either asexually, sexually, or both ways. Asexual reproduction occurs when a portion of a plant develops into a separate new plant. This new plant is genetically identical to the original, or parent, plant. Some plants, such as irises and daylilies, can use their underground stems for asexual reproduction. Other plants, such as the houseleeks, or hens and chicks, in **Figure 13,** reproduce asexually using horizontal stems called stolons. One advantage of asexual reproduction is that just one parent organism can produce offspring. However, sexual reproduction in plants usually requires two parent organisms. Sexual reproduction occurs when a plant's sperm combines with a plant's egg. A resulting zygote can grow into a plant. This new plant is a genetic combination of its parents.

✓ **Reading Check** How are sexual and asexual reproduction different in plants?

Figure 13 Hens and chicks can reproduce without seeds, or asexually. New "chicks" can grow from the stolons on the main "hen" plant.

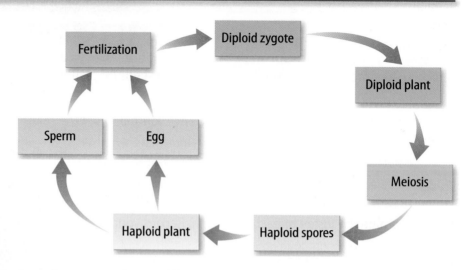

Figure 14 The life cycle of all plants includes an alternation of generations. The diploid generation begins with fertilization. The haploid generation begins with meiosis.

Alternation of Generations

Your body is made of two types of cells—haploid cells and diploid cells. Most of your cells are diploid. The only human haploid cells are sperm and eggs. As a result, you will live your entire life as a diploid organism. To put it another way, your life cycle includes only a diploid stage. That isn't true for all organisms. Some organisms, including plants, have two life stages called **generations.** One generation is almost all diploid cells. The other generation has only haploid cells. **Alternation of generations** *occurs when the life cycle of an organism alternates between diploid and haploid generations,* as shown in **Figure 14.**

 Key Concept Check What is alternation of generations in plants?

The Diploid Generation

When you look at a tree or a flower, you're seeing part of the plant's diploid generation. Meiosis occurs in certain cells in the reproductive structures of a diploid plant. *The daughter cells produced from haploid structures are called* **spores.** Spores grow by mitosis and cell division and form the haploid generation of a plant.

The Haploid Generation

In most plants, the haploid generation is tiny and lives surrounded by special tissues of the diploid plant. In other plants, the haploid generation lives on its own. Certain reproductive cells of the haploid generation produce haploid sperm or eggs by mitosis and cell division. Fertilization takes place when a sperm and an egg fuse and form a diploid zygote. Through mitosis and cell division, the zygote grows into the diploid generation of a plant.

Reproduction in Seedless Plants

Not all plants grow from seeds. The first land plants to inhabit Earth probably were seedless plants—plants that grow from haploid spores, not from seeds. The mosses and ferns in **Figure 15** are examples of seedless plants found on Earth today.

Life Cycle of a Moss

The life cycle of a moss is typical for some seedless plants. The tiny, green moss plants that carpet rocks, bark, and soil in moist areas are haploid plants. These plants grow by **mitosis** and cell division from haploid spores produced by the diploid generation. They have male structures that produce sperm and female structures that produce eggs. Fertilization results in a diploid zygote that grows by mitosis and cell division into the diploid generation of moss, such as the one shown in **Figure 15.** A diploid moss is tiny and not easily seen.

REVIEW VOCABULARY

mitosis
the process during which a nucleus and its contents divide

Life Cycle of a Fern

An alternation of generations is also seen in the life cycle of a fern. The diploid generations are the green leafy plants often seen in forests. These plants produce haploid spores. The spores grow into tiny plants. The haploid plants produce eggs and sperm that can unite and form the diploid generations.

 Key Concept Check How do seedless plants such as mosses and ferns reproduce?

Figure 15 Mosses and ferns usually grow in moist environments. Sperm must swim through a film of water to reach an egg.

Reproduction in Seedless Plants

Fern

Diploid generation of moss

Moss covering log

Concepts in Motion Animation

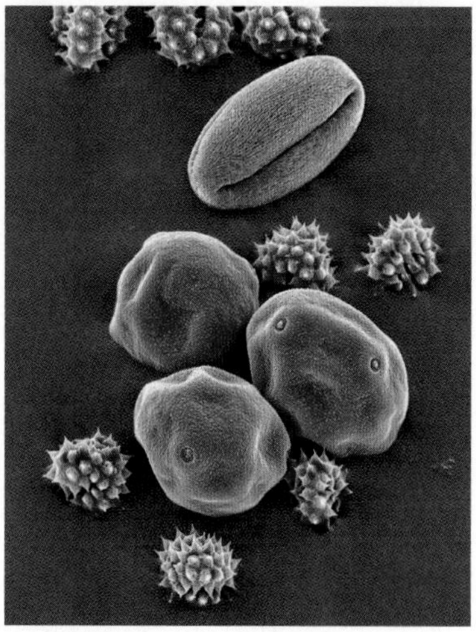

How do seed plants reproduce?

Most land plants that cover Earth grow from seeds. There are two groups of seed plants—flowerless seed plants and flowering seed plants.

Unlike seedless plants, the haploid generation of a seed plant is within diploid tissue. Separate diploid male and diploid female reproductive structures produce haploid sperm and haploid eggs that join during fertilization.

The Role of Pollen Grains

A **pollen** (PAH lun) **grain** *forms from tissue in a male reproductive structure of a seed plant.* Each pollen grain contains nutrients and has a hard, protective outer covering, as shown in **Figure 16.** Pollen grains produce sperm cells. Wind, animals, gravity, or water currents can carry pollen grains to female reproductive structures.

Plants cannot move and find a mate as most animals can. Do you recall reading about the yellow dust at the beginning of this lesson? That dust is pollen grains. Male reproductive structures produce a vast number of pollen grains. **Pollination** (pah luh NAY shun) *occurs when pollen grains land on a female reproductive structure of a plant that is the same species as the pollen grains.*

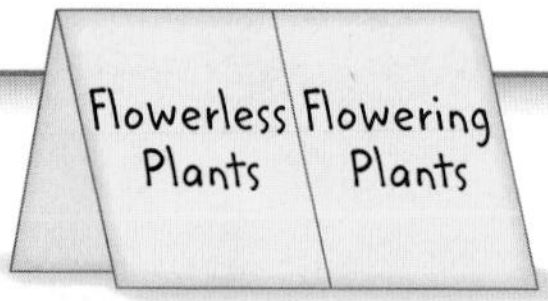

The Role of Ovules and Seeds

The female reproductive structure of a seed plant where the haploid egg develops is called the ovule. Following pollination, sperm enter the ovule and fertilization occurs. A zygote forms and develops into an embryo, *an immature diploid plant that develops from the zygote.* As shown in **Figure 17,** *an embryo, its food supply, and a protective covering make up a* seed. A seed's food supply provides the embryo with nourishment for its early growth.

 Key Concept Check How do seed plants reproduce?

Color-enhanced SEM Magnification: 1,100×

▲ **Figure 16** Pollen grains of one type of plant are different from those of any other type of plant.

◉ **Visual Check** How many different types of pollen are visible in **Figure 16?**

Figure 17 A seed contains a diploid plant embryo and a food supply protected by a hard outer covering.

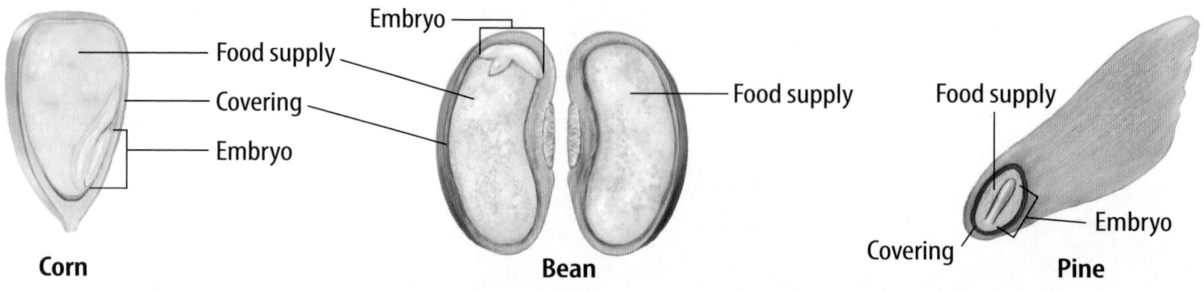

Corn
— Food supply
— Covering
— Embryo

Embryo
— Food supply
Bean

Food supply
Covering
Embryo
Pine

Reproduction in Flowerless Seed Plants

Flowerless seed plants are also known as gymnosperms (JIHM nuh spurmz). The word *gymnosperm* means "naked seed," and gymnosperm seeds are not surrounded by a fruit. The most common gymnosperms are conifers. Conifers, such as pines, firs, cypresses, redwoods, and yews, are trees and shrubs with needlelike or scalelike leaves. Most conifers are evergreens, which means they have leaves all year long. Conifers can live for many years. Bristlecone pines, such as the one shown in **Figure 18,** are among the oldest living trees on Earth.

Life Cycle of a Gymnosperm The life cycle of a gymnosperm, shown in **Figure 19,** includes an alternation of generations. Cones are the male and female reproductive structures of conifers. They contain the haploid generation. Male cones are small, papery structures that produce pollen grains. Female cones can be woody, berrylike, or soft, and they produce eggs. A zygote forms when a sperm from a male cone fertilizes an egg. The zygote is the beginning of the diploid generation. Seeds form as part of the female cone.

 Reading Check Where is the haploid generation of conifers contained?

▲ **Figure 18** Seeds form at the base of each scale on a female cone.

Reproduction in Flowerless Seed Plants 🔑

Concepts in Motion Animation

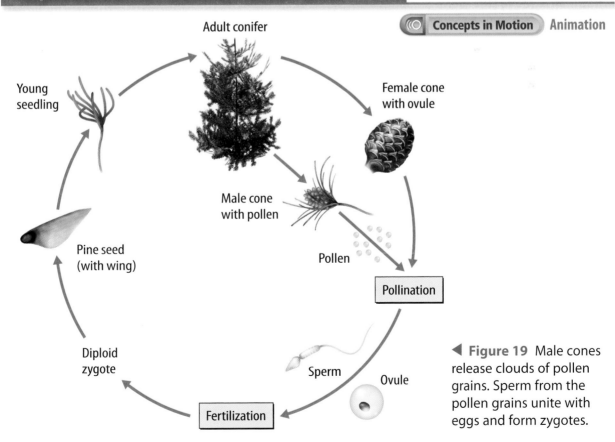

◀ **Figure 19** Male cones release clouds of pollen grains. Sperm from the pollen grains unite with eggs and form zygotes.

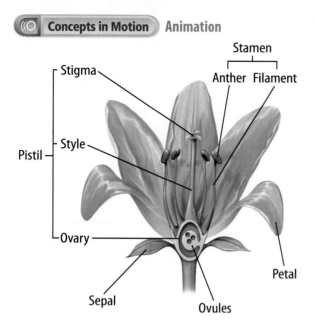

Figure 20 Typical flowers have both male and female structures.

Reproduction in Flowering Seed Plants

Most plants you see around you are angiosperms, or flowering plants. Fruits and vegetables come from angiosperms. Many animals depend on angiosperms for food.

The Flower Reproduction of an angiosperm begins in a flower. Most flowers have male and female reproductive structures, as shown in **Figure 20.**

The male reproductive organ of a flower is the **stamen.** Pollen grains form at the tip of the stamen in the anther. The filament supports the anther and connects it to the base of the flower. *The female reproductive organ of a flower is the* **pistil.** Pollen can land at the tip of the pistil, or stigma. The stigma is at the top of a long tube called the style. *At the base of the style is the* **ovary,** *which contains one or more ovules.* Recall that each ovule eventually will contain a haploid egg and might become a seed if fertilized.

(Inquiry) **MiniLab**

30 minutes

How can you model a flower?

Imagine that you have just discovered a new species of flowering plant. No one has ever seen this flower before, but it has all the basic flower parts.

1 Read and complete a lab safety form.

2 In your Science Journal, list all the parts your flower has as an angiosperm.

3 Make a large 3-dimensional model of your new flower using **chenille stems, tissue paper, construction paper, tag board, pom poms, plastic beads, scissors,** and **glue.**

4 Check your model to make sure each flower part is in the correct proportion and shows how it interacts with other flower parts.

5 Name your flower. Create a key to identify each part and its function.

Analyze and Conclude

1. **Analyze** Why do flowers have colorful petals and strong scents?

2. **Infer** Why does the end of the stigma feel sticky?

3. ⊙━ **Key Concept** Could your flower be self-pollinating? Explain.

Life Cycle of an Angiosperm A typical life cycle for an angiosperm is shown in **Figure 21.** Pollen grains travel by wind, gravity, water, or animal from the anther to the stigma, where pollination occurs. A pollen tube grows from the pollen grain into the stigma, down the style, to the ovary at the base of the pistil. Sperm develop from a haploid cell in the pollen tube. When the pollen tube enters an ovule, fertilization takes place.

As you read earlier, the zygote that results from fertilization develops into an embryo. Each ovule and its embryo will become a seed. *The ovary, and sometimes other parts of the flower, will develop into a* **fruit** *that contains one or more seeds.* The seeds can grow into new, genetically related plants that produce flowers, and the cycle repeats.

✓ **Reading Check** Do sperm develop before or after pollination?

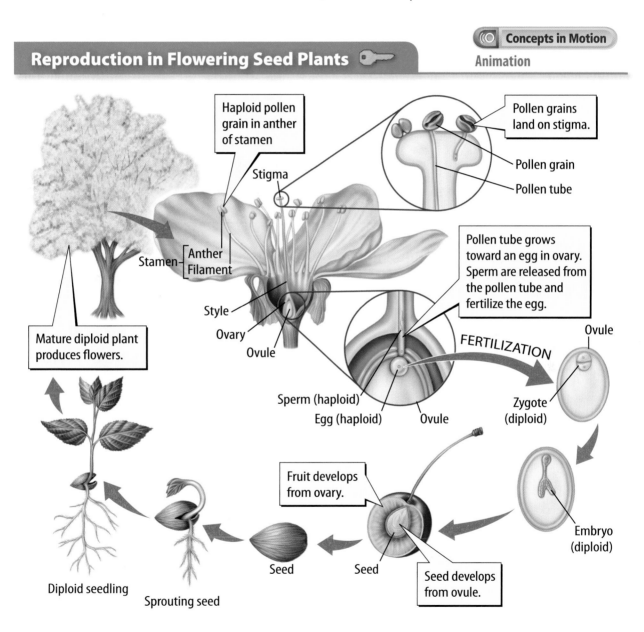

Reproduction in Flowering Seed Plants ⚷

((◎ **Concepts in Motion**
Animation

Haploid pollen grain in anther of stamen

Stigma

Pollen grains land on stigma.

Pollen grain

Pollen tube

Stamen— Anther Filament

Pollen tube grows toward an egg in ovary. Sperm are released from the pollen tube and fertilize the egg.

Style

Ovary

Ovule

Mature diploid plant produces flowers.

FERTILIZATION

Ovule

Sperm (haploid)

Egg (haploid)

Ovule

Zygote (diploid)

Embryo (diploid)

Fruit develops from ovary.

Seed develops from ovule.

Seed

Seed

Diploid seedling

Sprouting seed

Figure 21 During the life cycle of a flowering plant, the haploid generation grows and develops inside the diploid plant.

✓ **Visual Check** How does sperm in a pollen grain reach an egg in the ovule?

Table 1 Flowers, Fruits, and Seeds of Common Plants

Plant	Flower	Fruit	Seed
Pea			
Corn			
Strawberry			
Dandelion			

Table 1 Flowers, fruits, and seeds are important for reproduction in angiosperms.

Visual Check Which of these fruits has seeds on the outside?

Figure 22 The seeds will be excreted by the mouse and might grow into new blackberry bushes.

Fruit and Seed Dispersal Fruits and seeds, such as those in **Table 1,** are important sources of food for people and animals. In most cases, seeds of flowering plants are inside fruits. Pods are the fruits of a pea plant. The peas inside a pod are the seeds. An ear of corn is made up of many fruits, or kernels. The main part of each kernel is the seed. Strawberries have tiny seeds on the outside of the fruit.

Reading Check Where are the seeds of flowering plants usually found?

We usually think of fruits as juicy and edible, such as oranges or watermelons. However, some fruits are hard, dry, and not particularly edible. For example, each parachutelike structure of a dandelion is a dry fruit.

Fruits help protect seeds and disperse them. Some fruits, such as those of a dandelion, are light and float on air currents, which helps to scatter the seeds. Also, when an animal eats a fruit, the fruit's seeds can pass through the animal's digestive system with little or no damage to the seed. Imagine what happens when an animal, such as the mouse shown in **Figure 22,** eats blackberries. The animal digests the juicy fruit and deposits the seeds with its wastes. By this time, the animal might have traveled some distance away from the blackberry bush. This means the animal helped to disperse the seeds away from the blackberry bush.

Lesson 3 Review

Visual Summary

The life cycle of a plant includes an alternation of generations.

Seedless plants, such as ferns and mosses, grow from haploid spores.

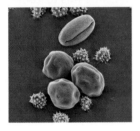

In seed plants, pollination occurs when pollen grains land on the female reproductive structure of a plant of the same species.

FOLDABLES®

Use your lesson Foldable to review the lesson. Save your Foldable for the project at the end of the chapter.

What do you think NOW?

You first read the statements below at the beginning of the chapter.

5. Seeds contain tiny plant embryos.

6. Flowers are needed for plant reproduction.

Did you change your mind about whether you agree or disagree with the statements? Rewrite any false statements to make them true.

Use Vocabulary

1 The daughter cells produced from haploid structures are called _____.

2 **Distinguish** between an ovule and an ovary.

3 **Define** *pollination* in your own words.

Understand Key Concepts

4 Which is NOT part of the alternation of generations life cycle in plants?
A. anther   C. haploid
B. diploid   D. spore

5 **Contrast** the haploid generation of a moss with that of a fern.

6 **Describe** how a pollen tube carries sperm to the ovule in a flower.

7 **Give an example** of a flowerless seed plant.

Interpret Graphics

8 **Examine** the figure below and describe the function of each part of the seed.

9 **Identify** Copy and fill in the graphic organizer below to identify the female parts of a flower.

Critical Thinking

10 **Create** a picture to show the life cycle of a fern.

11 **Evaluate** the advantages of fruit production in plant reproduction.

Materials

one quad of plants

Also needed
appropriate materials to perform lab

Safety

Design a Stimulating Environment for Plants

Plants usually respond to stimuli in the environment by growing. The response to light is phototropism; plants grow toward the light. The growth response of gravitropism is a little more complicated; stems grow away from the direction of gravity (negative gravitropism), and roots grow in the direction of gravity (positive gravitropism). Thigmotropism is a plant response to touch.

Ask a Question

You have explored tropisms in other labs in this chapter. What questions would you like to answer more thoroughly, or what outcomes would you like to double-check? Do you have another approach in mind to investigate one of the tropisms? Ask a question that you would like to investigate further. Make sure it is testable; think about the variables and equipment you would need.

Make Observations

1. Read and complete a lab safety form.
2. Examine your quad of plants and decide which tropism you want to explore.
3. Make a plan and write it in your Science Journal.
4. Have your teacher approve your plan for your investigation.

5. Choose materials from those provided by your teacher for a simple lab setup.
6. Decide the criteria you will use to show the outcomes you expect.
7. Set up your lab according to your plan.

Form a Hypothesis

8. After observing your plants and lab setup, formulate a hypothesis about the relationship between your selected tropism and a plant's growth. Make a prediction about how the tropism will affect your plants.

Test Your Hypothesis

9 Make any necessary modifications to your setup so your procedure will move toward your expected outcome.

10 Make a data table like the one to the right in your Science Journal.

11 Make your observations as directed by your procedure, and record them in your table.

Analyze and Conclude

12 **Compare** the position of the parts of your plant at the beginning and end of your study. Check to see if the change is easily visible and measurable; try not to jump to conclusions.

13 **Consider** the possible causes of the changes. Determine if it was changing the variable that brought about the effect. Explain.

14 **Relate** how the tropism you modeled could enable plants to meet their needs and survive.

15 🔵 **The Big Idea** What might happen if the stimulus you provided for the plant was enlarged, minimized, or eliminated?

Communicate Your Results

Prepare a drama to present your findings. Group members or volunteers from the class can wear pictures or signs to indicate their roles. Begin with students role-playing a healthy plant. Add the role of the stimulus, and be sure to identify the tropism and show results in the plant(s).

Inquiry **Extension**

Phototropism is one of the plant responses to stimuli that you have been able to explore easily by changing the position of the light source or the position of the plants in relation to the light source. What might happen if you changed the light source itself? What if you put a colored plastic sheet between the light and the plant? Would a red filter cause the same response as a green filter? What if you used different plants? For example, some mustard seeds are fast-germinating. Would these respond the same way as the other plants?

Time Period	[Variable observed] of Plant			
	1	2	3	4
Day 0 prior to tropism				
Day 1				
Day 2				
Day 3				

Lab Tips

☑ Discuss the possible materials you will use with your lab partner. Remember that the materials should help you learn more about the tropism you selected.

☑ Be creative when deciding how to test the tropism you selected.

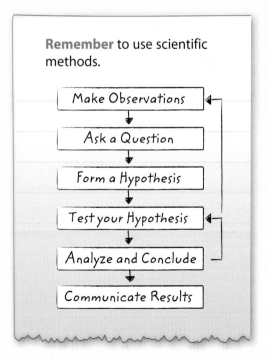

Remember to use scientific methods.

Make Observations → Ask a Question → Form a Hypothesis → Test your Hypothesis → Analyze and Conclude → Communicate Results

Chapter 10 Study Guide

Plants transform light energy into chemical energy, respond to stimuli and maintain homeostasis, and reproduce with and without seeds.

Key Concepts Summary 🔑

Lesson 1: Energy Processing in Plants

- The vascular tissue in most plants, xylem and phloem, move materials throughout plants.

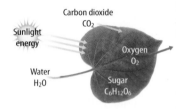

- In **photosynthesis,** plants convert light energy, water, and carbon dioxide into the food-energy molecule glucose through a series of chemical reactions. The process gives off oxygen.

- **Cellular respiration** is a series of chemical reactions that convert the energy in food molecules into a usable form of energy called ATP.

- Photosynthesis and cellular respiration can be considered opposite processes of each other.

Lesson 2: Plant Responses

- Although plants cannot move from one place to another, they do respond to **stimuli,** or changes in their environments. Plants respond to stimuli in different ways.

- **Tropisms** are growth responses toward or away from stimuli such as light, touch, and gravity. **Photoperiodism** is a plant's response to the number of hours of darkness in its environment.

- Plants respond to chemical stimuli, or **plant hormones,** such as auxins, ethylene, gibberellins, and cytokinins. Different hormones have different effects on plants.

Lesson 3: Plant Reproduction

- **Alternation of generations** is when the life cycle of an organism alternates between diploid and haploid generations.

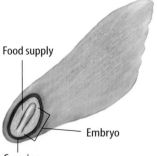

- Seedless plants, such as ferns, reproduce when a haploid sperm fertilizes a haploid egg, forming a diploid zygote.

- Seed plants reproduce when **pollen grains,** which contain haploid sperm, land on the tip of the female reproductive organ. At the base of this organ is the **ovary,** which usually contains one or more **ovules.** Each ovule eventually will contain a haploid egg. If the sperm fertilizes the egg, an **embryo** will form within a **seed.**

Vocabulary

Lesson 1:
photosynthesis p. 334
cellular respiration p. 336

Lesson 2:
stimulus p. 341
tropism p. 342
photoperiodism p. 344
plant hormone p. 345

Lesson 3:
alternation of generations p. 352
spore p. 352
pollen grain p. 354
pollination p. 354
ovule p. 354
embryo p. 354
seed p. 354
stamen p. 356
pistil p. 356
ovary p. 356
fruit p. 357

FOLDABLES® **Chapter Project**

Assemble your lesson Foldables as shown to make a Chapter Project. Use the project to review what you have learned in this chapter.

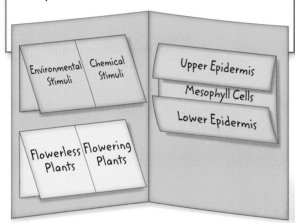

Environmental Stimuli Chemical Stimuli

Upper Epidermis

Mesophyll Cells

Lower Epidermis

Flowerless Plants Flowering Plants

Use Vocabulary

1. Long-day and short-day plants are examples of plants that respond to _____.

2. The process that uses oxygen and produces carbon dioxide is _____.

3. A(n) _____ forms from tissue in a male reproductive structure of a seed plant.

4. A(n) _____ develops from an ovary and surrounding tissue.

5. Sperm travel down the _____ inside the stigma of a flower to reach the ovary.

Link Vocabulary and Key Concepts

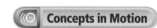 **Concepts in Motion** Interactive Concept Map

Copy this concept map, and then use vocabulary terms from the previous page to complete the concept map.

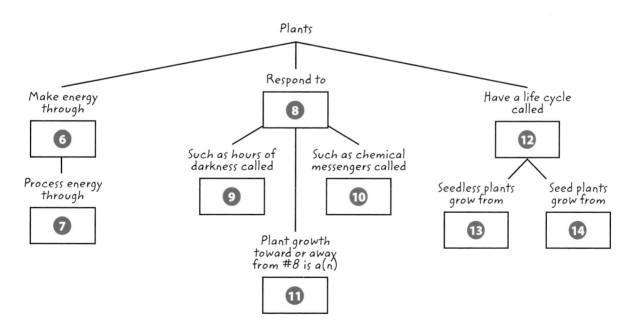

Plants

Make energy through

6

Process energy through

7

Respond to

8

Such as hours of darkness called

9

Such as chemical messengers called

10

Plant growth toward or away from #8 is a(n)

11

Have a life cycle called

12

Seedless plants grow from

13

Seed plants grow from

14

Chapter 10 Review

Understand Key Concepts 🔑

1 Which material travels from the roots to the leaves through the xylem?

A. oxygen
B. sugar
C. sunlight
D. water

2 Which organelle is the site of photosynthesis?

A. chloroplast
B. mitochondria
C. nucleus
D. ribosome

3 Which is a product of cellular respiration?

A. ATP
B. light
C. oxygen
D. sugar

Use the image below to answer questions 4 and 5.

4 What type of plant-growth response is shown in the photo above?

A. flowering
B. gravitropism
C. photoperiodism
D. thigmotropism

5 Which stimulus is responsible for this type of growth?

A. gravity
B. light
C. nutrients
D. touch

Use the image below to answer questions 6–8.

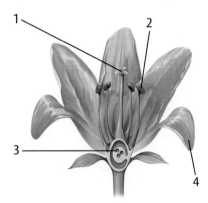

6 What is the name of structure number 3?

A. anther
B. ovule
C. petal
D. pistil

7 Where is pollen produced?

A. 1
B. 2
C. 3
D. 4

8 What part of a flower becomes a seed?

A. 1
B. 2
C. 3
D. 4

9 What plant is shown in the figure below?

A. diploid fern
B. diploid moss
C. haploid fern
D. haploid moss

Critical Thinking

10 Infer which came first—photosynthesis or cellular respiration.

11 Assess the importance of material transport in plants.

12 Construct a table to compare the reactants and products of photosynthesis and cellular respiration.

13 Evaluate the internal structure of a leaf as a location for photosynthesis.

14 Assess the need for plants to respond to their environment.

15 Predict what would happen if a short-day plant were exposed to more hours of daylight.

16 Critique the saying "one rotten apple spoils the whole barrel."

17 Infer from the photo below where the light source is in relation to the plant.

18 Evaluate the importance of fruit production in flowering plants.

19 Predict the effect of cold temperature killing all the flowers on fruit trees.

Writing in Science

20 Write a five-sentence paragraph about the importance of plants in your life. Include a main idea, supporting details, and a concluding sentence.

REVIEW THE BIG IDEA

21 Make a list of the plant processes you learned about in this chapter. How do these processes help a plant survive and reproduce?

22 How does the process shown below help a plant survive?

Math Skills

Review — Math Practice

Use Percentages

23 Without treatment with gibberellins, 500 out of 1,000 grass seeds germinated. When sprayed with gibberellins, 875 of the seeds germinated. What was the percentage increase?

24 A bunch of bananas ripens (turns from green to yellow) in 42 hours. When the bananas are placed in a bag with an apple, which releases ethylene, the bananas ripen in 21 hours. What is the percentage change in ripening time?

Standardized Test Practice

Record your answers on the answer sheet provided by your teacher or on a sheet of paper.

Multiple Choice

1 Which structure transports sugars throughout a plant?

  A epidermis

  B phloem

  C stomata

  D xylem

2 What is one similarity between plants and animals?

  A Both plants and animals carry on cellular respiration.

  B Both plants and animals carry on photosynthesis.

  C Both plants and animals have chloroplasts.

  D Both plants and animals use xylem and phloem to transport materials.

Use the diagram below to answer question 3.

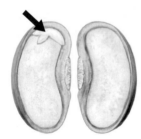

3 Look at the structure that is marked with an arrow in the image above. What will this structure become?

  A a diploid moss

  B a diploid seed plant

  C a haploid fern

  D a haploid flowerless seed plant

4 Which two plant hormones increase the rate of cell division?

  A auxins and cytokinins

  B cytokinins and giberellins

  C ethylene and auxins

  D giberellin and ethylene

5 Which is a product of photosynthesis?

  A carbon dioxide

  B glucose

  C light

  D water

Use the image below to answer question 6.

6 Which cellular process occurs within the organelle shown above?

  A photosynthesis

  B cellular respiration

  C transport of phloem

  D transport of xylem

7 Which plant has a diploid stage that is difficult to see?

  A conifer

  B cherry tree

  C dandelion

  D moss

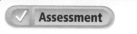
8 How is cellular respiration related to photosynthesis?

A Animals produce sugars through cellular respiration that are broken down by plants through photosynthesis.

B Animals use cellular respiration while plants use photosynthesis.

C Cellular respiration produces sugars, which are stored through photosynthesis.

D Photosynthesis produces sugars, which are broken down in cellular respiration.

Use the diagram below to answer question 9.

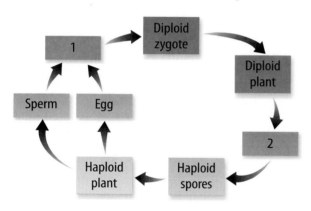

9 Which process occurs at the stage marked *2* on the plant life cycle diagram above?

A asexual reproduction

B fertilization

C meiosis

D mitosis

Constructed Response

Use the figure below to answer questions 10 and 11.

10 Describe what is happening in the image above. In your response, identify the environmental stimulus and the plant growth response.

11 Which plant hormone is involved in the growth response shown in the drawing above? Explain how this hormone causes the growth response.

12 Fruit distributors might use technologies to remove ethylene from the area where fruits are stored. How does this practice affect the fruit? Explain your answer.

13 Plants might reproduce both sexually and asexually. How are these two forms of reproduction similar? How are they different?

NEED EXTRA HELP?													
If You Missed Question...	1	2	3	4	5	6	7	8	9	10	11	12	13
Go to Lesson...	1	1	3	2	1	1	3	1	3	2	2	2	3

Student Resources

For Students and Parents/Guardians

These resources are designed to help you achieve success in science. You will find useful information on laboratory safety, math skills, and science skills. In addition, science reference materials are found in the Reference Handbook. You'll find the information you need to learn and sharpen your skills in these resources.

Table of Contents

Science Skill Handbook SR-2

Scientific Methods **SR-2**
Identify a Question.. SR-2
Gather and Organize Information............................ SR-2
Form a Hypothesis SR-5
Test the Hypothesis SR-6
Collect Data SR-6
Analyze the Data... SR-9
Draw Conclustions SR-10
Communicate.. SR-10
Safety Symbols **SR-11**
Safety in the Science Laboratory **SR-12**
General Safety Rules SR-12
Prevent Accidents... SR-12
Laboratory Work SR-13
Emergencies.. SR-13

Math Skill Handbook SR-14

Math Review. **SR-14**
Use Fractions SR-14
Use Ratios SR-17
Use Decimals SR-17
Use Proportions.. SR-18
Use Percentages SR-19
Solve One-Step Equations SR-19
Use Statistics... SR-20
Use Geometry SR-21
Science Application **SR-24**
Measure in SI SR-24
Dimensional Analysis...................................... SR-24
Precision and Significant Digits SR-26
Scientific Notation SR-26
Make and Use Graphs SR-27

Foldables Handbook SR-29

Reference Handbook SR-40
Periodic Table of the Elements............................. SR-40
Diversity of Life: Classification of Living Organisms SR-42
Use and Care of a Microscope SR-46

Glossary G-2

Index I-2

Credits C-2

Scientific Methods

Scientists use an orderly approach called the scientific method to solve problems. This includes organizing and recording data so others can understand them. Scientists use many variations in this method when they solve problems.

Identify a Question

The first step in a scientific investigation or experiment is to identify a question to be answered or a problem to be solved. For example, you might ask which gasoline is the most efficient.

Gather and Organize Information

After you have identified your question, begin gathering and organizing information. There are many ways to gather information, such as researching in a library, interviewing those knowledgeable about the subject, and testing and working in the laboratory and field. Fieldwork is investigations and observations done outside of a laboratory.

Researching Information Before moving in a new direction, it is important to gather the information that already is known about the subject. Start by asking yourself questions to determine exactly what you need to know. Then you will look for the information in various reference sources, like the student is doing in **Figure 1.** Some sources may include textbooks, encyclopedias, government documents, professional journals, science magazines, and the Internet. Always list the sources of your information.

Figure 1 The Internet can be a valuable research tool.

Evaluate Sources of Information Not all sources of information are reliable. You should evaluate all of your sources of information, and use only those you know to be dependable. For example, if you are researching ways to make homes more energy efficient, a site written by the U.S. Department of Energy would be more reliable than a site written by a company that is trying to sell a new type of weatherproofing material. Also, remember that research always is changing. Consult the most current resources available to you. For example, a 1985 resource about saving energy would not reflect the most recent findings.

Sometimes scientists use data that they did not collect themselves, or conclusions drawn by other researchers. This data must be evaluated carefully. Ask questions about how the data were obtained, if the investigation was carried out properly, and if it has been duplicated exactly with the same results. Would you reach the same conclusion from the data? Only when you have confidence in the data can you believe it is true and feel comfortable using it.

SCIENCE SKILL HANDBOOK

MATH SKILL HANDBOOK

FOLDABLES HANDBOOK

REFERENCE HANDBOOK

GLOSSARY/ GLOSARIO

INDEX

Interpret Scientific Illustrations As you research a topic in science, you will see drawings, diagrams, and photographs to help you understand what you read. Some illustrations are included to help you understand an idea that you can't see easily by yourself, like the tiny particles in an atom in **Figure 2.** A drawing helps many people to remember details more easily and provides examples that clarify difficult concepts or give additional information about the topic you are studying. Most illustrations have labels or a caption to identify or to provide more information.

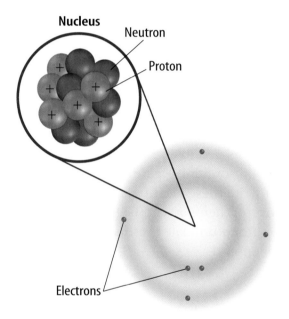

Figure 2 This drawing shows an atom of carbon with its six protons, six neutrons, and six electrons.

Concept Maps One way to organize data is to draw a diagram that shows relationships among ideas (or concepts). A concept map can help make the meanings of ideas and terms more clear, and help you understand and remember what you are studying. Concept maps are useful for breaking large concepts down into smaller parts, making learning easier.

Network Tree A type of concept map that not only shows a relationship, but how the concepts are related is a network tree, shown in **Figure 3.** In a network tree, the words are written in the ovals, while the description of the type of relationship is written across the connecting lines.

When constructing a network tree, write down the topic and all major topics on separate pieces of paper or notecards. Then arrange them in order from general to specific. Branch the related concepts from the major concept and describe the relationship on the connecting line. Continue to more specific concepts until finished.

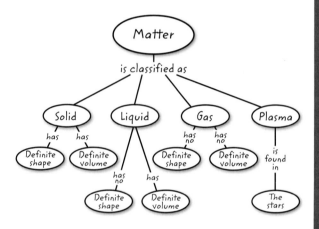

Figure 3 A network tree shows how concepts or objects are related.

Events Chain Another type of concept map is an events chain. Sometimes called a flow chart, it models the order or sequence of items. An events chain can be used to describe a sequence of events, the steps in a procedure, or the stages of a process.

When making an events chain, first find the one event that starts the chain. This event is called the initiating event. Then, find the next event and continue until the outcome is reached, as shown in **Figure 4** on the next page.

SCIENCE SKILL HANDBOOK

MATH SKILL HANDBOOK

FOLDABLES HANDBOOK

REFERENCE HANDBOOK

GLOSSARY/ GLOSARIO

INDEX

SCIENCE SKILL HANDBOOK

MATH SKILL HANDBOOK

FOLDABLES HANDBOOK

REFERENCE HANDBOOK

GLOSSARY/ GLOSARIO

INDEX

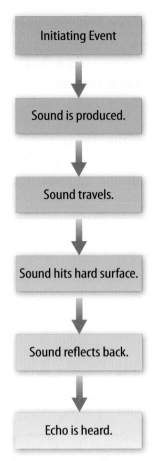

Figure 4 Events-chain concept maps show the order of steps in a process or event. This concept map shows how a sound makes an echo.

Cycle Map A specific type of events chain is a cycle map. It is used when the series of events do not produce a final outcome, but instead relate back to the beginning event, such as in **Figure 5.** Therefore, the cycle repeats itself.

To make a cycle map, first decide what event is the beginning event. This is also called the initiating event. Then list the next events in the order that they occur, with the last event relating back to the initiating event. Words can be written between the events that describe what happens from one event to the next. The number of events in a cycle map can vary, but usually contain three or more events.

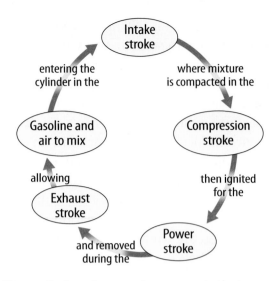

Figure 5 A cycle map shows events that occur in a cycle.

Spider Map A type of concept map that you can use for brainstorming is the spider map. When you have a central idea, you might find that you have a jumble of ideas that relate to it but are not necessarily clearly related to each other. The spider map on sound in **Figure 6** shows that if you write these ideas outside the main concept, then you can begin to separate and group unrelated terms so they become more useful.

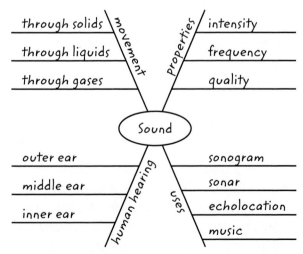

Figure 6 A spider map allows you to list ideas that relate to a central topic but not necessarily to one another.

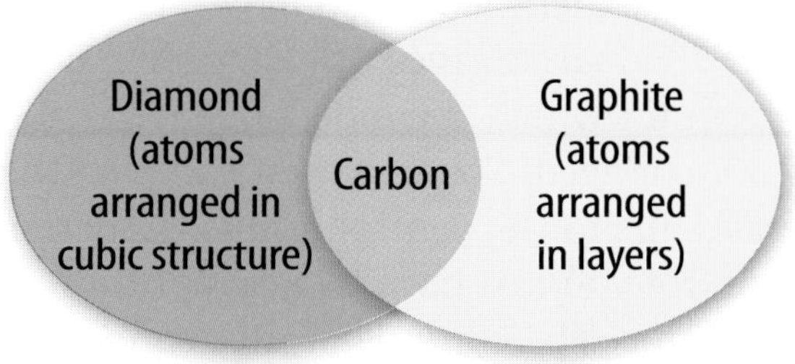

Figure 7 This Venn diagram compares and contrasts two substances made from carbon.

Venn Diagram To illustrate how two subjects compare and contrast you can use a Venn diagram. You can see the characteristics that the subjects have in common and those that they do not, shown in **Figure 7.**

To create a Venn diagram, draw two overlapping ovals that are big enough to write in. List the characteristics unique to one subject in one oval, and the characteristics of the other subject in the other oval. The characteristics in common are listed in the overlapping section.

Make and Use Tables One way to organize information so it is easier to understand is to use a table. Tables can contain numbers, words, or both.

To make a table, list the items to be compared in the first column and the characteristics to be compared in the first row. The title should clearly indicate the content of the table, and the column or row heads should be clear. Notice that in **Table 1** the units are included.

Table 1 Recyclables Collected During Week			
Day of Week	Paper (kg)	Aluminum (kg)	Glass (kg)
Monday	5.0	4.0	12.0
Wednesday	4.0	1.0	10.0
Friday	2.5	2.0	10.0

Make a Model One way to help you better understand the parts of a structure, the way a process works, or to show things too large or small for viewing is to make a model. For example, an atomic model made of a plastic-ball nucleus and chenille stem electron shells can help you visualize how the parts of an atom relate to each other. Other types of models can be devised on a computer or represented by equations.

Form a Hypothesis

A possible explanation based on previous knowledge and observations is called a hypothesis. After researching gasoline types and recalling previous experiences in your family's car you form a hypothesis—our car runs more efficiently because we use premium gasoline. To be valid, a hypothesis has to be something you can test by using an investigation.

Predict When you apply a hypothesis to a specific situation, you predict something about that situation. A prediction makes a statement in advance, based on prior observation, experience, or scientific reasoning. People use predictions to make everyday decisions. Scientists test predictions by performing investigations. Based on previous observations and experiences, you might form a prediction that cars are more efficient with premium gasoline. The prediction can be tested in an investigation.

Design an Experiment A scientist needs to make many decisions before beginning an investigation. Some of these include: how to carry out the investigation, what steps to follow, how to record the data, and how the investigation will answer the question. It also is important to address any safety concerns.

SCIENCE SKILL HANDBOOK

MATH SKILL HANDBOOK

FOLDABLES HANDBOOK

REFERENCE HANDBOOK

GLOSSARY/ GLOSARIO

INDEX

SCIENCE SKILL HANDBOOK

MATH SKILL HANDBOOK

FOLDABLES HANDBOOK

REFERENCE HANDBOOK

GLOSSARY/ GLOSARIO

INDEX

Test the Hypothesis

Now that you have formed your hypothesis, you need to test it. Using an investigation, you will make observations and collect data, or information. This data might either support or not support your hypothesis. Scientists collect and organize data as numbers and descriptions.

Follow a Procedure In order to know what materials to use, as well as how and in what order to use them, you must follow a procedure. **Figure 8** shows a procedure you might follow to test your hypothesis.

Procedure

Step 1	Use regular gasoline for two weeks.
Step 2	Record the number of kilometers between fill-ups and the amount of gasoline used.
Step 3	Switch to premium gasoline for two weeks.
Step 4	Record the number of kilometers between fill-ups and the amount of gasoline used.

Figure 8 A procedure tells you what to do step-by-step.

Identify and Manipulate Variables and Controls In any experiment, it is important to keep everything the same except for the item you are testing. The one factor you change is called the independent variable. The change that results is the dependent variable. Make sure you have only one independent variable, to assure yourself of the cause of the changes you observe in the dependent variable. For example, in your gasoline experiment the type of fuel is the independent variable. The dependent variable is the efficiency.

Many experiments also have a control—an individual instance or experimental subject for which the independent variable is not changed. You can then compare the test results to the control results. To design a control you can have two cars of the same type. The control car uses regular gasoline for four weeks. After you are done with the test, you can compare the experimental results to the control results.

Collect Data

Whether you are carrying out an investigation or a short observational experiment, you will collect data, as shown in **Figure 9.** Scientists collect data as numbers and descriptions and organize them in specific ways.

Observe Scientists observe items and events, then record what they see. When they use only words to describe an observation, it is called qualitative data. Scientists' observations also can describe how much there is of something. These observations use numbers, as well as words, in the description and are called quantitative data. For example, if a sample of the element gold is described as being "shiny and very dense" the data are qualitative. Quantitative data on this sample of gold might include "a mass of 30 g and a density of 19.3 g/cm^3."

Figure 9 Collecting data is one way to gather information directly.

Figure 10 Record data neatly and clearly so it is easy to understand.

When you make observations you should examine the entire object or situation first, and then look carefully for details. It is important to record observations accurately and completely. Always record your notes immediately as you make them, so you do not miss details or make a mistake when recording results from memory. Never put unidentified observations on scraps of paper. Instead they should be recorded in a notebook, like the one in **Figure 10.** Write your data neatly so you can easily read it later. At each point in the experiment, record your observations and label them. That way, you will not have to determine what the figures mean when you look at your notes later. Set up any tables that you will need to use ahead of time, so you can record any observations right away. Remember to avoid bias when collecting data by not including personal thoughts when you record observations. Record only what you observe.

Estimate Scientific work also involves estimating. To estimate is to make a judgment about the size or the number of something without measuring or counting. This is important when the number or size of an object or population is too large or too difficult to accurately count or measure.

Sample Scientists may use a sample or a portion of the total number as a type of estimation. To sample is to take a small, representative portion of the objects or organisms of a population for research. By making careful observations or manipulating variables within that portion of the group, information is discovered and conclusions are drawn that might apply to the whole population. A poorly chosen sample can be unrepresentative of the whole. If you were trying to determine the rainfall in an area, it would not be best to take a rainfall sample from under a tree.

Measure You use measurements every day. Scientists also take measurements when collecting data. When taking measurements, it is important to know how to use measuring tools properly. Accuracy also is important.

Length To measure length, the distance between two points, scientists use meters. Smaller measurements might be measured in centimeters or millimeters.

Length is measured using a metric ruler or meterstick. When using a metric ruler, line up the 0-cm mark with the end of the object being measured and read the number of the unit where the object ends. Look at the metric ruler shown in **Figure 11.** The centimeter lines are the long, numbered lines, and the shorter lines are millimeter lines. In this instance, the length would be 4.50 cm.

Figure 11 This metric ruler has centimeter and millimeter divisions.

SCIENCE SKILL HANDBOOK

MATH SKILL HANDBOOK

FOLDABLES HANDBOOK

REFERENCE HANDBOOK

GLOSSARY/ GLOSARIO

INDEX

SCIENCE SKILL HANDBOOK

MATH SKILL HANDBOOK

FOLDABLES HANDBOOK

REFERENCE HANDBOOK

GLOSSARY/ GLOSARIO

INDEX

Mass The SI unit for mass is the kilogram (kg). Scientists can measure mass using units formed by adding metric prefixes to the unit gram (g), such as milligram (mg). To measure mass, you might use a triple-beam balance similar to the one shown in **Figure 12.** The balance has a pan on one side and a set of beams on the other side. Each beam has a rider that slides on the beam.

When using a triple-beam balance, place an object on the pan. Slide the largest rider along its beam until the pointer drops below zero. Then move it back one notch. Repeat the process for each rider proceeding from the larger to smaller until the pointer swings an equal distance above and below the zero point. Sum the masses on each beam to find the mass of the object. Move all riders back to zero when finished.

Instead of putting materials directly on the balance, scientists often take a tare of a container. A tare is the mass of a container into which objects or substances are placed for measuring their masses. To find the mass of objects or substances, find the mass of a clean container. Remove the container from the pan, and place the object or substances in the container. Find the mass of the container with the materials in it. Subtract the mass of the empty container from the mass of the filled container to find the mass of the materials you are using.

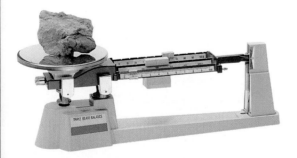

Figure 12 A triple-beam balance is used to determine the mass of an object.

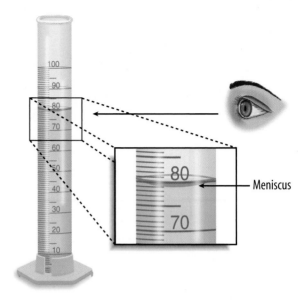

Figure 13 Graduated cylinders measure liquid volume.

Liquid Volume To measure liquids, the unit used is the liter. When a smaller unit is needed, scientists might use a milliliter. Because a milliliter takes up the volume of a cube measuring 1 cm on each side it also can be called a cubic centimeter ($cm^3 = cm \times cm \times cm$).

You can use beakers and graduated cylinders to measure liquid volume. A graduated cylinder, shown in **Figure 13,** is marked from bottom to top in milliliters. In lab, you might use a 10-mL graduated cylinder or a 100-mL graduated cylinder. When measuring liquids, notice that the liquid has a curved surface. Look at the surface at eye level, and measure the bottom of the curve. This is called the meniscus. The graduated cylinder in **Figure 13** contains 79.0 mL, or 79.0 cm^3, of a liquid.

Temperature Scientists often measure temperature using the Celsius scale. Pure water has a freezing point of 0°C and boiling point of 100°C. The unit of measurement is degrees Celsius. Two other scales often used are the Fahrenheit and Kelvin scales.

Figure 14 A thermometer measures the temperature of an object.

Scientists use a thermometer to measure temperature. Most thermometers in a laboratory are glass tubes with a bulb at the bottom end containing a liquid such as colored alcohol. The liquid rises or falls with a change in temperature. To read a glass thermometer like the thermometer in **Figure 14,** rotate it slowly until a red line appears. Read the temperature where the red line ends.

Form Operational Definitions An operational definition defines an object by how it functions, works, or behaves. For example, when you are playing hide and seek and a tree is home base, you have created an operational definition for a tree.

Objects can have more than one operational definition. For example, a ruler can be defined as a tool that measures the length of an object (how it is used). It can also be a tool with a series of marks used as a standard when measuring (how it works).

Analyze the Data

To determine the meaning of your observations and investigation results, you will need to look for patterns in the data. Then you must think critically to determine what the data mean. Scientists use several approaches when they analyze the data they have collected and recorded. Each approach is useful for identifying specific patterns.

Interpret Data The word *interpret* means "to explain the meaning of something." When analyzing data from an experiment, try to find out what the data show. Identify the control group and the test group to see whether changes in the independent variable have had an effect. Look for differences in the dependent variable between the control and test groups.

Classify Sorting objects or events into groups based on common features is called classifying. When classifying, first observe the objects or events to be classified. Then select one feature that is shared by some members in the group, but not by all. Place those members that share that feature in a subgroup. You can classify members into smaller and smaller subgroups based on characteristics. Remember that when you classify, you are grouping objects or events for a purpose. Keep your purpose in mind as you select the features to form groups and subgroups.

Compare and Contrast Observations can be analyzed by noting the similarities and differences between two or more objects or events that you observe. When you look at objects or events to see how they are similar, you are comparing them. Contrasting is looking for differences in objects or events.

SCIENCE SKILL HANDBOOK

MATH SKILL HANDBOOK

FOLDABLES HANDBOOK

REFERENCE HANDBOOK

GLOSSARY/ GLOSARIO

INDEX

Recognize Cause and Effect A cause is a reason for an action or condition. The effect is that action or condition. When two events happen together, it is not necessarily true that one event caused the other. Scientists must design a controlled investigation to recognize the exact cause and effect.

Draw Conclusions

When scientists have analyzed the data they collected, they proceed to draw conclusions about the data. These conclusions are sometimes stated in words similar to the hypothesis that you formed earlier. They may confirm a hypothesis, or lead you to a new hypothesis.

Infer Scientists often make inferences based on their observations. An inference is an attempt to explain observations or to indicate a cause. An inference is not a fact, but a logical conclusion that needs further investigation. For example, you may infer that a fire has caused smoke. Until you investigate, however, you do not know for sure.

Apply When you draw a conclusion, you must apply those conclusions to determine whether the data supports the hypothesis. If your data do not support your hypothesis, it does not mean that the hypothesis is wrong. It means only that the result of the investigation did not support the hypothesis. Maybe the experiment needs to be redesigned, or some of the initial observations on which the hypothesis was based were incomplete or biased. Perhaps more observation or research is needed to refine your hypothesis. A successful investigation does not always come out the way you originally predicted.

Avoid Bias Sometimes a scientific investigation involves making judgments. When you make a judgment, you form an opinion. It is important to be honest and not to allow any expectations of results to bias your judgments. This is important throughout the entire investigation, from researching to collecting data to drawing conclusions.

Communicate

The communication of ideas is an important part of the work of scientists. A discovery that is not reported will not advance the scientific community's understanding or knowledge. Communication among scientists also is important as a way of improving their investigations.

Scientists communicate in many ways, from writing articles in journals and magazines that explain their investigations and experiments, to announcing important discoveries on television and radio. Scientists also share ideas with colleagues on the Internet or present them as lectures, like the student is doing in **Figure 15.**

Figure 15 A student communicates to his peers about his investigation.

SCIENCE SKILL HANDBOOK

MATH SKILL HANDBOOK

FOLDABLES HANDBOOK

REFERENCE HANDBOOK

GLOSSARY/ GLOSARIO

INDEX

These safety symbols are used in laboratory and field investigations in this book to indicate possible hazards. Learn the meaning of each symbol and refer to this page often. *Remember to wash your hands thoroughly after completing lab procedures.*

PROTECTIVE EQUIPMENT   Do not begin any lab without the proper protection equipment.

GOGGLES	Proper eye protection must be worn when performing or observing science activities that involve items or conditions as listed below.	**APRON** Wear an approved apron when using substances that could stain, wet, or destroy cloth.	**SOAP** Wash hands with soap and water before removing goggles and after all lab activities.	**GLOVES** Wear gloves when working with biological materials, chemicals, animals, or materials that can stain or irritate hands.

LABORATORY HAZARDS

Symbols	Potential Hazards	Precaution	Response
DISPOSAL	contamination of classroom or environment due to improper disposal of materials such as chemicals and live specimens	• DO NOT dispose of hazardous materials in the sink or trash can. • Dispose of wastes as directed by your teacher.	• If hazardous materials are disposed of improperly, notify your teacher immediately.
EXTREME TEMPERATURE	skin burns due to extremely hot or cold materials such as hot glass, liquids, or metals; liquid nitrogen; dry ice	• Use proper protective equipment, such as hot mitts and/or tongs, when handling objects with extreme temperatures.	• If injury occurs, notify your teacher immediately.
SHARP OBJECTS	punctures or cuts from sharp objects such as razor blades, pins, scalpels, and broken glass	• Handle glassware carefully to avoid breakage. • Walk with sharp objects pointed downward, away from you and others.	• If broken glass or injury occurs, notify your teacher immediately.
ELECTRICAL	electric shock or skin burn due to improper grounding, short circuits, liquid spills, or exposed wires	• Check condition of wires and apparatus for fraying or uninsulated wires, and broken or cracked equipment. • Use only GFCI-protected outlets	• DO NOT attempt to fix electrical problems. Notify your teacher immediately.
CHEMICAL	skin irritation or burns, breathing difficulty, and/or poisoning due to touching, swallowing, or inhalation of chemicals such as acids, bases, bleach, metal compounds, iodine, poinsettias, pollen, ammonia, acetone, nail polish remover, heated chemicals, mothballs, and any other chemicals labeled or known to be dangerous	• Wear proper protective equipment such as goggles, apron, and gloves when using chemicals. • Ensure proper room ventilation or use a fume hood when using materials that produce fumes. • NEVER smell fumes directly. • NEVER taste or eat any material in the laboratory.	• If contact occurs, immediately flush affected area with water and notify your teacher. • If a spill occurs, leave the area immediately and notify your teacher.
FLAMMABLE	unexpected fire due to liquids or gases that ignite easily such as rubbing alcohol	• Avoid open flames, sparks, or heat when flammable liquids are present.	• If a fire occurs, leave the area immediately and notify your teacher.
OPEN FLAME	burns or fire due to open flame from matches, Bunsen burners, or burning materials	• Tie back loose hair and clothing. • Keep flame away from all materials. • Follow teacher instructions when lighting and extinguishing flames. • Use proper protection, such as hot mitts or tongs, when handling hot objects.	• If a fire occurs, leave the area immediately and notify your teacher.
ANIMAL SAFETY	injury to or from laboratory animals	• Wear proper protective equipment such as gloves, apron, and goggles when working with animals. • Wash hands after handling animals.	• If injury occurs, notify your teacher immediately.
BIOLOGICAL	infection or adverse reaction due to contact with organisms such as bacteria, fungi, and biological materials such as blood, animal or plant materials	• Wear proper protective equipment such as gloves, goggles, and apron when working with biological materials. • Avoid skin contact with an organism or any part of the organism. • Wash hands after handling organisms.	• If contact occurs, wash the affected area and notify your teacher immediately.
FUME	breathing difficulties from inhalation of fumes from substances such as ammonia, acetone, nail polish remover, heated chemicals, and mothballs	• Wear goggles, apron, and gloves. • Ensure proper room ventilation or use a fume hood when using substances that produce fumes. • NEVER smell fumes directly.	• If a spill occurs, leave area and notify your teacher immediately.
IRRITANT	irritation of skin, mucous membranes, or respiratory tract due to materials such as acids, bases, bleach, pollen, mothballs, steel wool, and potassium permanganate	• Wear goggles, apron, and gloves. • Wear a dust mask to protect against fine particles.	• If skin contact occurs, immediately flush the affected area with water and notify your teacher.
RADIOACTIVE	excessive exposure from alpha, beta, and gamma particles	• Remove gloves and wash hands with soap and water before removing remainder of protective equipment.	• If cracks or holes are found in the container, notify your teacher immediately.

Safety in the Science Laboratory

Introduction to Science Safety

The science laboratory is a safe place to work if you follow standard safety procedures. Being responsible for your own safety helps to make the entire laboratory a safer place for everyone. When performing any lab, read and apply the caution statements and safety symbol listed at the beginning of the lab.

General Safety Rules

1. Complete the *Lab Safety Form* or other safety contract BEFORE starting any science lab.

2. Study the procedure. Ask your teacher any questions. Be sure you understand safety symbols shown on the page.

3. Notify your teacher about allergies or other health conditions that can affect your participation in a lab.

4. Learn and follow use and safety procedures for your equipment. If unsure, ask your teacher.

5. Never eat, drink, chew gum, apply cosmetics, or do any personal grooming in the lab. Never use lab glassware as food or drink containers. Keep your hands away from your face and mouth.

6. Know the location and proper use of the safety shower, eye wash, fire blanket, and fire alarm.

Prevent Accidents

1. Use the safety equipment provided to you. Goggles and a safety apron should be worn during investigations.

2. Do NOT use hair spray, mousse, or other flammable hair products. Tie back long hair and tie down loose clothing.

3. Do NOT wear sandals or other open-toed shoes in the lab.

4. Remove jewelry on hands and wrists. Loose jewelry, such as chains and long necklaces, should be removed to prevent them from getting caught in equipment.

5. Do not taste any substances or draw any material into a tube with your mouth.

6. Proper behavior is expected in the lab. Practical jokes and fooling around can lead to accidents and injury.

7. Keep your work area uncluttered.

Laboratory Work

1. Collect and carry all equipment and materials to your work area before beginning a lab.

2. Remain in your own work area unless given permission by your teacher to leave it.

SCIENCE SKILL HANDBOOK

MATH SKILL HANDBOOK

FOLDABLES HANDBOOK

REFERENCE HANDBOOK

GLOSSARY/ GLOSARIO

INDEX

3. Always slant test tubes away from yourself and others when heating them, adding substances to them, or rinsing them.

4. If instructed to smell a substance in a container, hold the container a short distance away and fan vapors toward your nose.

5. Do NOT substitute other chemicals/substances for those in the materials list unless instructed to do so by your teacher.

6. Do NOT take any materials or chemicals outside of the laboratory.

7. Stay out of storage areas unless instructed to be there and supervised by your teacher.

Laboratory Cleanup

1. Turn off all burners, water, and gas, and disconnect all electrical devices.

2. Clean all pieces of equipment and return all materials to their proper places.

3. Dispose of chemicals and other materials as directed by your teacher. Place broken glass and solid substances in the proper containers. Never discard materials in the sink.

4. Clean your work area.

5. Wash your hands with soap and water thoroughly BEFORE removing your goggles.

Emergencies

1. Report any fire, electrical shock, glassware breakage, spill, or injury, no matter how small, to your teacher immediately. Follow his or her instructions.

2. If your clothing should catch fire, STOP, DROP, and ROLL. If possible, smother it with the fire blanket or get under a safety shower. NEVER RUN.

3. If a fire should occur, turn off all gas and leave the room according to established procedures.

4. In most instances, your teacher will clean up spills. Do NOT attempt to clean up spills unless you are given permission and instructions to do so.

5. If chemicals come into contact with your eyes or skin, notify your teacher immediately. Use the eyewash, or flush your skin or eyes with large quantities of water.

6. The fire extinguisher and first-aid kit should only be used by your teacher unless it is an extreme emergency and you have been given permission.

7. If someone is injured or becomes ill, only a professional medical provider or someone certified in first aid should perform first-aid procedures.

SCIENCE SKILL HANDBOOK

MATH SKILL HANDBOOK

FOLDABLES HANDBOOK

REFERENCE HANDBOOK

GLOSSARY/ GLOSARIO

INDEX

SCIENCE SKILL HANDBOOK

MATH SKILL HANDBOOK

FOLDABLES HANDBOOK

REFERENCE HANDBOOK

GLOSSARY/ GLOSARIO

INDEX

Use Fractions

A fraction compares a part to a whole. In the fraction $\frac{2}{3}$, the 2 represents the part and is the numerator. The 3 represents the whole and is the denominator.

Reduce Fractions To reduce a fraction, you must find the largest factor that is common to both the numerator and the denominator, the greatest common factor (GCF). Divide both numbers by the GCF. The fraction has then been reduced, or it is in its simplest form.

Example

Twelve of the 20 chemicals in the science lab are in powder form. What fraction of the chemicals used in the lab are in powder form?

Step 1 Write the fraction.

$$\frac{\text{part}}{\text{whole}} = \frac{12}{20}$$

Step 2 To find the GCF of the numerator and denominator, list all of the factors of each number.

Factors of 12: 1, 2, 3, 4, 6, 12 (the numbers that divide evenly into 12)

Factors of 20: 1, 2, 4, 5, 10, 20 (the numbers that divide evenly into 20)

Step 3 List the common factors.

1, 2, 4

Step 4 Choose the greatest factor in the list. The GCF of 12 and 20 is 4.

Step 5 Divide the numerator and denominator by the GCF.

$$\frac{12 \div 4}{20 \div 4} = \frac{3}{5}$$

In the lab, $\frac{3}{5}$ of the chemicals are in powder form.

Practice Problem At an amusement park, 66 of 90 rides have a height restriction. What fraction of the rides, in its simplest form, has a height restriction?

Add and Subtract Fractions with Like Denominators To add or subtract fractions with the same denominator, add or subtract the numerators and write the sum or difference over the denominator. After finding the sum or difference, find the simplest form for your fraction.

Example 1

In the forest outside your house, $\frac{1}{8}$ of the animals are rabbits, $\frac{3}{8}$ are squirrels, and the remainder are birds and insects. How many are mammals?

Step 1 Add the numerators.

$$\frac{1}{8} + \frac{3}{8} = \frac{(1 + 3)}{8} = \frac{4}{8}$$

Step 2 Find the GCF.

$$\frac{4}{8} \text{ (GCF, 4)}$$

Step 3 Divide the numerator and denominator by the GCF.

$$\frac{4 \div 4}{8 \div 4} = \frac{1}{2}$$

$\frac{1}{2}$ of the animals are mammals.

Example 2

If $\frac{7}{16}$ of the Earth is covered by freshwater, and $\frac{1}{16}$ of that is in glaciers, how much freshwater is not frozen?

Step 1 Subtract the numerators.

$$\frac{7}{16} - \frac{1}{16} = \frac{(7 - 1)}{16} = \frac{6}{16}$$

Step 2 Find the GCF.

$$\frac{6}{16} \text{ (GCF, 2)}$$

Step 3 Divide the numerator and denominator by the GCF.

$$\frac{6 \div 2}{16 \div 2} = \frac{3}{8}$$

$\frac{3}{8}$ of the freshwater is not frozen.

Practice Problem A bicycle rider is riding at a rate of 15 km/h for $\frac{4}{9}$ of his ride, 10 km/h for $\frac{2}{9}$ of his ride, and 8 km/h for the remainder of the ride. How much of his ride is he riding at a rate greater than 8 km/h?

Add and Subtract Fractions with Unlike Denominators To add or subtract fractions with unlike denominators, first find the least common denominator (LCD). This is the smallest number that is a common multiple of both denominators. Rename each fraction with the LCD, and then add or subtract. Find the simplest form if necessary.

Example 1

A chemist makes a paste that is $\frac{1}{2}$ table salt (NaCl), $\frac{1}{3}$ sugar ($C_6H_{12}O_6$), and the remainder is water (H_2O). How much of the paste is a solid?

Step 1 Find the LCD of the fractions.

$\frac{1}{2} + \frac{1}{3}$ (LCD, 6)

Step 2 Rename each numerator and each denominator with the LCD.

Step 3 Add the numerators.

$\frac{3}{6} + \frac{2}{6} = \frac{(3+2)}{6} = \frac{5}{6}$

$\frac{5}{6}$ of the paste is a solid.

Example 2

The average precipitation in Grand Junction, CO, is $\frac{7}{10}$ inch in November, and $\frac{3}{5}$ inch in December. What is the total average precipitation?

Step 1 Find the LCD of the fractions.

$\frac{7}{10} + \frac{3}{5}$ (LCD, 10)

Step 2 Rename each numerator and each denominator with the LCD.

Step 3 Add the numerators.

$\frac{7}{10} + \frac{6}{10} = \frac{(7+6)}{10} = \frac{13}{10}$

$\frac{13}{10}$ inches total precipitation, or $1\frac{3}{10}$ inches.

Practice Problem On an electric bill, about $\frac{1}{8}$ of the energy is from solar energy and about $\frac{1}{10}$ is from wind power. How much of the total bill is from solar energy and wind power combined?

Example 3

In your body, $\frac{7}{10}$ of your muscle contractions are involuntary (cardiac and smooth muscle tissue). Smooth muscle makes $\frac{3}{15}$ of your muscle contractions. How many of your muscle contractions are made by cardiac muscle?

Step 1 Find the LCD of the fractions.

$\frac{7}{10} - \frac{3}{15}$ (LCD, 30)

Step 2 Rename each numerator and each denominator with the LCD.

$\frac{7 \times 3}{10 \times 3} = \frac{21}{30}$

$\frac{3 \times 2}{15 \times 2} = \frac{6}{30}$

Step 3 Subtract the numerators.

$\frac{21}{30} - \frac{6}{30} = \frac{(21-6)}{30} = \frac{15}{30}$

Step 4 Find the GCF.

$\frac{15}{30}$ (GCF, 15)

$\frac{1}{2}$

$\frac{1}{2}$ of all muscle contractions are cardiac muscle.

Example 4

Tony wants to make cookies that call for $\frac{3}{4}$ of a cup of flour, but he only has $\frac{1}{3}$ of a cup. How much more flour does he need?

Step 1 Find the LCD of the fractions.

$\frac{3}{4} - \frac{1}{3}$ (LCD, 12)

Step 2 Rename each numerator and each denominator with the LCD.

$\frac{3 \times 3}{4 \times 3} = \frac{9}{12}$

$\frac{1 \times 4}{3 \times 4} = \frac{4}{12}$

Step 3 Subtract the numerators.

$\frac{9}{12} - \frac{4}{12} = \frac{(9-4)}{12} = \frac{5}{12}$

$\frac{5}{12}$ of a cup of flour

Practice Problem Using the information provided to you in Example 3 above, determine how many muscle contractions are voluntary (skeletal muscle).

SCIENCE SKILL HANDBOOK

MATH SKILL HANDBOOK

FOLDABLES HANDBOOK

REFERENCE HANDBOOK

GLOSSARY/ GLOSARIO

INDEX

Multiply Fractions To multiply with fractions, multiply the numerators and multiply the denominators. Find the simplest form if necessary.

Example

Multiply $\frac{3}{5}$ by $\frac{1}{3}$.

Step 1 Multiply the numerators and denominators.

$$\frac{3}{5} \times \frac{1}{3} = \frac{(3 \times 1)}{(5 \times 3)} \; \frac{3}{15}$$

Step 2 Find the GCF.

$$\frac{3}{15} \;(GCF, 3)$$

Step 3 Divide the numerator and denominator by the GCF.

$$\frac{3 \div 3}{15 \div 3} = \frac{1}{5}$$

$\frac{3}{5}$ multiplied by $\frac{1}{3}$ is $\frac{1}{5}$.

Practice Problem Multiply $\frac{3}{14}$ by $\frac{5}{16}$.

Find a Reciprocal Two numbers whose product is 1 are called multiplicative inverses, or reciprocals.

Example

Find the reciprocal of $\frac{3}{8}$.

Step 1 Inverse the fraction by putting the denominator on top and the numerator on the bottom.

$$\frac{8}{3}$$

The reciprocal of $\frac{3}{8}$ is $\frac{8}{3}$.

Practice Problem Find the reciprocal of $\frac{4}{9}$.

Divide Fractions To divide one fraction by another fraction, multiply the dividend by the reciprocal of the divisor. Find the simplest form if necessary.

Example 1

Divide $\frac{1}{9}$ by $\frac{1}{3}$.

Step 1 Find the reciprocal of the divisor.

The reciprocal of $\frac{1}{3}$ is $\frac{3}{1}$.

Step 2 Multiply the dividend by the reciprocal of the divisor.

$$\frac{\frac{1}{9}}{\frac{1}{3}} = \frac{1}{9} \times \frac{3}{1} = \frac{(1 \times 3)}{(9 \times 1)} = \frac{3}{9}$$

Step 3 Find the GCF.

$$\frac{3}{9} \;(GCF, 3)$$

Step 4 Divide the numerator and denominator by the GCF.

$$\frac{3 \div 3}{9 \div 3} = \frac{1}{3}$$

$\frac{1}{9}$ divided by $\frac{1}{3}$ is $\frac{1}{3}$.

Example 2

Divide $\frac{3}{5}$ by $\frac{1}{4}$.

Step 1 Find the reciprocal of the divisor.

The reciprocal of $\frac{1}{4}$ is $\frac{4}{1}$.

Step 2 Multiply the dividend by the reciprocal of the divisor.

$$\frac{\frac{3}{5}}{\frac{1}{4}} = \frac{3}{5} \times \frac{4}{1} = \frac{(3 \times 4)}{(5 \times 1)} = \frac{12}{5}$$

$\frac{3}{5}$ divided by $\frac{1}{4}$ is $\frac{12}{5}$ or $2\frac{2}{5}$.

Practice Problem Divide $\frac{3}{11}$ by $\frac{7}{10}$.

SCIENCE SKILL HANDBOOK

MATH SKILL HANDBOOK

FOLDABLES HANDBOOK

REFERENCE HANDBOOK

GLOSSARY/ GLOSARIO

INDEX

Use Ratios

When you compare two numbers by division, you are using a ratio. Ratios can be written 3 to 5, 3:5, or $\frac{3}{5}$. Ratios, like fractions, also can be written in simplest form.

Ratios can represent one type of probability, called odds. This is a ratio that compares the number of ways a certain outcome occurs to the number of possible outcomes. For example, if you flip a coin 100 times, what are the odds that it will come up heads? There are two possible outcomes, heads or tails, so the odds of coming up heads are 50:100. Another way to say this is that 50 out of 100 times the coin will come up heads. In its simplest form, the ratio is 1:2.

Example 1

A chemical solution contains 40 g of salt and 64 g of baking soda. What is the ratio of salt to baking soda as a fraction in simplest form?

Step 1 Write the ratio as a fraction.

$$\frac{\text{salt}}{\text{baking soda}} = \frac{40}{64}$$

Step 2 Express the fraction in simplest form. The GCF of 40 and 64 is 8.

$$\frac{40}{64} = \frac{40 \div 8}{64 \div 8} = \frac{5}{8}$$

The ratio of salt to baking soda in the sample is 5:8.

Example 2

Sean rolls a 6-sided die 6 times. What are the odds that the side with a 3 will show?

Step 1 Write the ratio as a fraction.

$$\frac{\text{number of sides with a 3}}{\text{number of possible sides}} = \frac{1}{6}$$

Step 2 Multiply by the number of attempts.

$$\frac{1}{6} \times 6 \text{ attempts} = \frac{6}{6} \text{ attempts} = 1 \text{ attempt}$$

1 attempt out of 6 will show a 3.

Practice Problem Two metal rods measure 100 cm and 144 cm in length. What is the ratio of their lengths in simplest form?

Use Decimals

A fraction with a denominator that is a power of ten can be written as a decimal. For example, 0.27 means $\frac{27}{100}$. The decimal point separates the ones place from the tenths place.

Any fraction can be written as a decimal using division. For example, the fraction $\frac{5}{8}$ can be written as a decimal by dividing 5 by 8. Written as a decimal, it is 0.625.

Add or Subtract Decimals When adding and subtracting decimals, line up the decimal points before carrying out the operation.

Example 1

Find the sum of 47.68 and 7.80.

Step 1 Line up the decimal places when you write the numbers.

$$\begin{array}{r} 47.68 \\ + 7.80 \end{array}$$

Step 2 Add the decimals.

$$\begin{array}{r} \overset{1\,1}{47.68} \\ + 7.80 \\ \hline 55.48 \end{array}$$

The sum of 47.68 and 7.80 is 55.48.

Example 2

Find the difference of 42.17 and 15.85.

Step 1 Line up the decimal places when you write the number.

$$\begin{array}{r} 42.17 \\ -15.85 \end{array}$$

Step 2 Subtract the decimals.

$$\begin{array}{r} \overset{3\,11}{4\!\!\!/2}.\overset{1}{1}7 \\ -15.85 \\ \hline 26.32 \end{array}$$

The difference of 42.17 and 15.85 is 26.32.

Practice Problem Find the sum of 1.245 and 3.842.

SCIENCE SKILL HANDBOOK

MATH SKILL HANDBOOK

FOLDABLES HANDBOOK

REFERENCE HANDBOOK

GLOSSARY/ GLOSARIO

INDEX

SCIENCE SKILL HANDBOOK

MATH SKILL HANDBOOK

FOLDABLES HANDBOOK

REFERENCE HANDBOOK

GLOSSARY/ GLOSARIO

INDEX

Multiply Decimals To multiply decimals, multiply the numbers like numbers without decimal points. Count the decimal places in each factor. The product will have the same number of decimal places as the sum of the decimal places in the factors.

Example

Multiply 2.4 by 5.9.

Step 1 Multiply the factors like two whole numbers.

$24 \times 59 = 1416$

Step 2 Find the sum of the number of decimal places in the factors. Each factor has one decimal place, for a sum of two decimal places.

Step 3 The product will have two decimal places.

14.16

The product of 2.4 and 5.9 is 14.16.

Practice Problem Multiply 4.6 by 2.2.

Divide Decimals When dividing decimals, change the divisor to a whole number. To do this, multiply both the divisor and the dividend by the same power of ten. Then place the decimal point in the quotient directly above the decimal point in the dividend. Then divide as you do with whole numbers.

Example

Divide 8.84 by 3.4.

Step 1 Multiply both factors by 10.

$3.4 \times 10 = 34, 8.84 \times 10 = 88.4$

Step 2 Divide 88.4 by 34.

```
        2.6
   34)88.4
      −68
       204
      −204
         0
```

8.84 divided by 3.4 is 2.6.

Practice Problem Divide 75.6 by 3.6.

Use Proportions

An equation that shows that two ratios are equivalent is a proportion. The ratios $\frac{2}{4}$ and $\frac{5}{10}$ are equivalent, so they can be written as $\frac{2}{4} = \frac{5}{10}$. This equation is a proportion.

When two ratios form a proportion, the cross products are equal. To find the cross products in the proportion $\frac{2}{4} = \frac{5}{10}$, multiply the 2 and the 10, and the 4 and the 5. Therefore $2 \times 10 = 4 \times 5$, or $20 = 20$.

Because you know that both ratios are equal, you can use cross products to find a missing term in a proportion. This is known as solving the proportion.

Example

The heights of a tree and a pole are proportional to the lengths of their shadows. The tree casts a shadow of 24 m when a 6-m pole casts a shadow of 4 m. What is the height of the tree?

Step 1 Write a proportion.

$$\frac{\text{height of tree}}{\text{height of pole}} = \frac{\text{length of tree's shadow}}{\text{length of pole's shadow}}$$

Step 2 Substitute the known values into the proportion. Let h represent the unknown value, the height of the tree.

$$\frac{h}{6} \times \frac{24}{4}$$

Step 3 Find the cross products.

$$h \times 4 = 6 \times 24$$

Step 4 Simplify the equation.

$$4h \times 144$$

Step 5 Divide each side by 4.

$$\frac{4h}{4} \times \frac{144}{4}$$

$$h = 36$$

The height of the tree is 36 m.

Practice Problem The ratios of the weights of two objects on the Moon and on Earth are in proportion. A rock weighing 3 N on the Moon weighs 18 N on Earth. How much would a rock that weighs 5 N on the Moon weigh on Earth?

Use Percentages

The word *percent* means "out of one hundred." It is a ratio that compares a number to 100. Suppose you read that 77 percent of Earth's surface is covered by water. That is the same as reading that the fraction of Earth's surface covered by water is $\frac{77}{100}$. To express a fraction as a percent, first find the equivalent decimal for the fraction. Then, multiply the decimal by 100 and add the percent symbol.

Example 1

Express $\frac{13}{20}$ as a percent.

Step 1 Find the equivalent decimal for the fraction.

$$
\begin{array}{r}
0.65 \\
20\overline{)13.00} \\
\underline{12\ 0} \\
1\ 00 \\
\underline{1\ 00} \\
0
\end{array}
$$

Step 2 Rewrite the fraction $\frac{13}{20}$ as 0.65.

Step 3 Multiply 0.65 by 100 and add the % symbol.

$$0.65 \times 100 = 65 = 65\%$$

So, $\frac{13}{20} = 65\%$.

This also can be solved as a proportion.

Example 2

Express $\frac{13}{20}$ as a percent.

Step 1 Write a proportion.

$$\frac{13}{20} = \frac{x}{100}$$

Step 2 Find the cross products.

$$1300 = 20x$$

Step 3 Divide each side by 20.

$$\frac{1300}{20} = \frac{20x}{20}$$

$$65\% = x$$

Practice Problem In one year, 73 of 365 days were rainy in one city. What percent of the days in that city were rainy?

Solve One-Step Equations

A statement that two expressions are equal is an equation. For example, $A = B$ is an equation that states that A is equal to B.

An equation is solved when a variable is replaced with a value that makes both sides of the equation equal. To make both sides equal the inverse operation is used. Addition and subtraction are inverses, and multiplication and division are inverses.

Example 1

Solve the equation $x - 10 = 35$.

Step 1 Find the solution by adding 10 to each side of the equation.

$$x - 10 = 35$$
$$x - 10 + 10 = 35 - 10$$
$$x = 45$$

Step 2 Check the solution.

$$x - 10 = 35$$
$$45 - 10 = 35$$
$$35 = 35$$

Both sides of the equation are equal, so $x = 45$.

Example 2

In the formula $a = bc$, find the value of c if $a = 20$ and $b = 2$.

Step 1 Rearrange the formula so the unknown value is by itself on one side of the equation by dividing both sides by b.

$$a = bc$$
$$\frac{a}{b} = \frac{bc}{b}$$
$$\frac{a}{b} = c$$

Step 2 Replace the variables a and b with the values that are given.

$$\frac{a}{b} = c$$
$$\frac{20}{2} = c$$
$$10 = c$$

Step 3 Check the solution.

$$a = bc$$
$$20 = 2 \times 10$$
$$20 = 20$$

Both sides of the equation are equal, so $c = 10$ is the solution when $a = 20$ and $b = 2$.

Practice Problem In the formula $h = gd$, find the value of d if $g = 12.3$ and $h = 17.4$.

SCIENCE SKILL HANDBOOK

MATH SKILL HANDBOOK

FOLDABLES HANDBOOK

REFERENCE HANDBOOK

GLOSSARY/ GLOSARIO

INDEX

SCIENCE SKILL HANDBOOK

MATH SKILL HANDBOOK

FOLDABLES HANDBOOK

REFERENCE HANDBOOK

GLOSSARY/ GLOSARIO

INDEX

Use Statistics

The branch of mathematics that deals with collecting, analyzing, and presenting data is statistics. In statistics, there are three common ways to summarize data with a single number—the mean, the median, and the mode.

The **mean** of a set of data is the arithmetic average. It is found by adding the numbers in the data set and dividing by the number of items in the set.

The **median** is the middle number in a set of data when the data are arranged in numerical order. If there were an even number of data points, the median would be the mean of the two middle numbers.

The **mode** of a set of data is the number or item that appears most often.

Another number that often is used to describe a set of data is the range. The **range** is the difference between the largest number and the smallest number in a set of data.

Example

The speeds (in m/s) for a race car during five different time trials are 39, 37, 44, 36, and 44.

To find the mean:

Step 1   Find the sum of the numbers.

$39 + 37 + 44 + 36 + 44 = 200$

Step 2   Divide the sum by the number of items, which is 5.

$200 \div 5 = 40$

The mean is 40 m/s.

To find the median:

Step 1   Arrange the measures from least to greatest.

36, 37, 39, 44, 44

Step 2   Determine the middle measure.

36, 37, <u>39</u>, 44, 44

The median is 39 m/s.

To find the mode:

Step 1   Group the numbers that are the same together.

44, 44, 36, 37, 39

Step 2   Determine the number that occurs most in the set.

<u>44, 44</u>, 36, 37, 39

The mode is 44 m/s.

To find the range:

Step 1   Arrange the measures from greatest to least.

44, 44, 39, 37, 36

Step 2   Determine the greatest and least measures in the set.

<u>44</u>, 44, 39, 37, 36

Step 3   Find the difference between the greatest and least measures.

$44 - 36 = 8$

The range is 8 m/s.

Practice Problem Find the mean, median, mode, and range for the data set 8, 4, 12, 8, 11, 14, 16.

A **frequency table** shows how many times each piece of data occurs, usually in a survey. **Table 1** below shows the results of a student survey on favorite color.

Table 1 Student Color Choice		
Color	**Tally**	**Frequency**
red	IIII	4
blue	⦀	5
black	II	2
green	III	3
purple	⦀ II	7
yellow	⦀ I	6

Based on the frequency table data, which color is the favorite?

Use Geometry

The branch of mathematics that deals with the measurement, properties, and relationships of points, lines, angles, surfaces, and solids is called geometry.

Perimeter The **perimeter** (P) is the distance around a geometric figure. To find the perimeter of a rectangle, add the length and width and multiply that sum by two, or $2(l + w)$. To find perimeters of irregular figures, add the length of the sides.

Example 1

Find the perimeter of a rectangle that is 3 m long and 5 m wide.

Step 1 You know that the perimeter is 2 times the sum of the width and length.

$P = 2(3 \text{ m} + 5 \text{ m})$

Step 2 Find the sum of the width and length.

$P = 2(8 \text{ m})$

Step 3 Multiply by 2.

$P = 16 \text{ m}$

The perimeter is 16 m.

Example 2

Find the perimeter of a shape with sides measuring 2 cm, 5 cm, 6 cm, 3 cm.

Step 1 You know that the perimeter is the sum of all the sides.

$P = 2 + 5 + 6 + 3$

Step 2 Find the sum of the sides.

$P = 2 + 5 + 6 + 3$

$P = 16$

The perimeter is 16 cm.

Practice Problem Find the perimeter of a rectangle with a length of 18 m and a width of 7 m.

Practice Problem Find the perimeter of a triangle measuring 1.6 cm by 2.4 cm by 2.4 cm.

Area of a Rectangle The **area** (A) is the number of square units needed to cover a surface. To find the area of a rectangle, multiply the length times the width, or $l \times w$. When finding area, the units also are multiplied. Area is given in square units.

Example

Find the area of a rectangle with a length of 1 cm and a width of 10 cm.

Step 1 You know that the area is the length multiplied by the width.

$A = (1 \text{ cm} \times 10 \text{ cm})$

Step 2 Multiply the length by the width. Also multiply the units.

$A = 10 \text{ cm}^2$

The area is 10 cm^2.

Practice Problem Find the area of a square whose sides measure 4 m.

Area of a Triangle To find the area of a triangle, use the formula:

$A = \frac{1}{2}(\text{base} \times \text{height})$

The base of a triangle can be any of its sides. The height is the perpendicular distance from a base to the opposite endpoint, or vertex.

Example

Find the area of a triangle with a base of 18 m and a height of 7 m.

Step 1 You know that the area is $\frac{1}{2}$ the base times the height.

$A = \frac{1}{2}(18 \text{ m} \times 7 \text{ m})$

Step 2 Multiply $\frac{1}{2}$ by the product of 18×7. Multiply the units.

$A = \frac{1}{2}(126 \text{ m}^2)$

$A = 63 \text{ m}^2$

The area is 63 m^2.

Practice Problem Find the area of a triangle with a base of 27 cm and a height of 17 cm.

SCIENCE SKILL HANDBOOK

MATH SKILL HANDBOOK

FOLDABLES HANDBOOK

REFERENCE HANDBOOK

GLOSSARY/ GLOSARIO

INDEX

Circumference of a Circle The **diameter** (*d*) of a circle is the distance across the circle through its center, and the **radius** (r) is the distance from the center to any point on the circle. The radius is half of the diameter. The distance around the circle is called the **circumference** (C). The formula for finding the circumference is:

$$C = 2\pi r \text{ or } C = \pi d$$

The circumference divided by the diameter is always equal to 3.1415926… This nonterminating and nonrepeating number is represented by the Greek letter π (pi). An approximation often used for π is 3.14.

Example 1

Find the circumference of a circle with a radius of 3 m.

Step 1 You know the formula for the circumference is 2 times the radius times π.

$$C = 2\pi(3)$$

Step 2 Multiply 2 times the radius.

$$C = 6\pi$$

Step 3 Multiply by π.

$$C \approx 19 \text{ m}$$

The circumference is about 19 m.

Example 2

Find the circumference of a circle with a diameter of 24.0 cm.

Step 1 You know the formula for the circumference is the diameter times π.

$$C = \pi(24.0)$$

Step 2 Multiply the diameter by π.

$$C \approx 75.4 \text{ cm}$$

The circumference is about 75.4 cm.

Practice Problem Find the circumference of a circle with a radius of 19 cm.

Area of a Circle The formula for the area of a circle is: $A = \pi r^2$

Example 1

Find the area of a circle with a radius of 4.0 cm.

Step 1 $A = \pi(4.0)^2$

Step 2 Find the square of the radius.

$$A = 16\pi$$

Step 3 Multiply the square of the radius by π.

$$A \approx 50 \text{ cm}^2$$

The area of the circle is about 50 cm².

Example 2

Find the area of a circle with a radius of 225 m.

Step 1 $A = \pi(225)^2$

Step 2 Find the square of the radius.

$$A = 50625\pi$$

Step 3 Multiply the square of the radius by π.

$$A \approx 159043.1$$

The area of the circle is about 159043.1 m².

Example 3

Find the area of a circle whose diameter is 20.0 mm.

Step 1 Remember that the radius is half of the diameter.

$$A = \pi\left(\frac{20.0}{2}\right)^2$$

Step 2 Find the radius.

$$A = \pi(10.0)^2$$

Step 3 Find the square of the radius.

$$A = 100\pi$$

Step 4 Multiply the square of the radius by π.

$$A \approx 314 \text{ mm}^2$$

The area of the circle is about 314 mm².

Practice Problem Find the area of a circle with a radius of 16 m.

SCIENCE SKILL HANDBOOK

MATH SKILL HANDBOOK

FOLDABLES HANDBOOK

REFERENCE HANDBOOK

GLOSSARY/ GLOSARIO

INDEX

Volume The measure of space occupied by a solid is the **volume** (V). To find the volume of a rectangular solid multiply the length times width times height, or $V = l \times w \times h$. It is measured in cubic units, such as cubic centimeters (cm^3).

Example

Find the volume of a rectangular solid with a length of 2.0 m, a width of 4.0 m, and a height of 3.0 m.

Step 1 You know the formula for volume is the length times the width times the height.

$V = 2.0\ m \times 4.0\ m \times 3.0\ m$

Step 2 Multiply the length times the width times the height.

$V = 24\ m^3$

The volume is 24 m^3.

Practice Problem Find the volume of a rectangular solid that is 8 m long, 4 m wide, and 4 m high.

To find the volume of other solids, multiply the area of the base times the height.

Example 1

Find the volume of a solid that has a triangular base with a length of 8.0 m and a height of 7.0 m. The height of the entire solid is 15.0 m.

Step 1 You know that the base is a triangle, and the area of a triangle is $\frac{1}{2}$ the base times the height, and the volume is the area of the base times the height.

$V = \left[\frac{1}{2}(b \times h)\right] \times 15$

Step 2 Find the area of the base.

$V = \left[\frac{1}{2}(8 \times 7)\right] \times 15$

$V = \left(\frac{1}{2} \times 56\right) \times 15$

Step 3 Multiply the area of the base by the height of the solid.

$V = 28 \times 15$

$V = 420\ m^3$

The volume is 420 m^3.

Example 2

Find the volume of a cylinder that has a base with a radius of 12.0 cm, and a height of 21.0 cm.

Step 1 You know that the base is a circle, and the area of a circle is the square of the radius times π, and the volume is the area of the base times the height.

$V = (\pi r^2) \times 21$

$V = (\pi 12^2) \times 21$

Step 2 Find the area of the base.

$V = 144\pi \times 21$

$V = 452 \times 21$

Step 3 Multiply the area of the base by the height of the solid.

$V \approx 9,500\ cm^3$

The volume is about 9,500 cm^3.

Example 3

Find the volume of a cylinder that has a diameter of 15 mm and a height of 4.8 mm.

Step 1 You know that the base is a circle with an area equal to the square of the radius times π. The radius is one-half the diameter. The volume is the area of the base times the height.

$V = (\pi r^2) \times 4.8$

$V = \left[\pi\left(\frac{1}{2} \times 15\right)^2\right] \times 4.8$

$V = (\pi 7.5^2) \times 4.8$

Step 2 Find the area of the base.

$V = 56.25\pi \times 4.8$

$V \approx 176.71 \times 4.8$

Step 3 Multiply the area of the base by the height of the solid.

$V \approx 848.2$

The volume is about 848.2 mm^3.

Practice Problem Find the volume of a cylinder with a diameter of 7 cm in the base and a height of 16 cm.

SCIENCE SKILL HANDBOOK

MATH SKILL HANDBOOK

FOLDABLES HANDBOOK

REFERENCE HANDBOOK

GLOSSARY/ GLOSARIO

INDEX

Science Applications

SCIENCE SKILL HANDBOOK

MATH SKILL HANDBOOK

FOLDABLES HANDBOOK

REFERENCE HANDBOOK

GLOSSARY/ GLOSARIO

INDEX

Measure in SI

The metric system of measurement was developed in 1795. A modern form of the metric system, called the International System (SI), was adopted in 1960 and provides the standard measurements that all scientists around the world can understand.

The SI system is convenient because unit sizes vary by powers of 10. Prefixes are used to name units. Look at **Table 2** for some common SI prefixes and their meanings.

Table 2 Common SI Prefixes

Prefix	Symbol	Meaning	
kilo–	k	1,000	thousandth
hecto–	h	100	hundred
deka–	da	10	ten
deci–	d	0.1	tenth
centi–	c	0.01	hundreth
milli–	m	0.001	thousandth

Example

How many grams equal one kilogram?

Step 1 Find the prefix *kilo–* in **Table 2.**

Step 2 Using **Table 2,** determine the meaning of *kilo–*. According to the table, it means 1,000. When the prefix *kilo–* is added to a unit, it means that there are 1,000 of the units in a "kilounit."

Step 3 Apply the prefix to the units in the question. The units in the question are grams. There are 1,000 grams in a kilogram.

Practice Problem Is a milligram larger or smaller than a gram? How many of the smaller units equal one larger unit? What fraction of the larger unit does one smaller unit represent?

Dimensional Analysis

Convert SI Units In science, quantities such as length, mass, and time sometimes are measured using different units. A process called dimensional analysis can be used to change one unit of measure to another. This process involves multiplying your starting quantity and units by one or more conversion factors. A conversion factor is a ratio equal to one and can be made from any two equal quantities with different units. If 1,000 mL equal 1 L then two ratios can be made.

$$\frac{1,000 \text{ mL}}{1 \text{ L}} = \frac{1 \text{ L}}{1,000 \text{ mL}} = 1$$

One can convert between units in the SI system by using the equivalents in **Table 2** to make conversion factors.

Example

How many cm are in 4 m?

Step 1 Write conversion factors for the units given. From **Table 2,** you know that 100 cm = 1 m. The conversion factors are

$$\frac{100 \text{ cm}}{1 \text{ m}} \text{ and } \frac{1 \text{ m}}{100 \text{ cm}}$$

Step 2 Decide which conversion factor to use. Select the factor that has the units you are converting from (m) in the denominator and the units you are converting to (cm) in the numerator.

$$\frac{100 \text{ cm}}{1 \text{ m}}$$

Step 3 Multiply the starting quantity and units by the conversion factor. Cancel the starting units with the units in the denominator. There are 400 cm in 4 m.

$$4 \text{ m} = \frac{100 \text{ cm}}{1 \text{ m}} = 400 \text{ cm}$$

Practice Problem How many milligrams are in one kilogram? (Hint: You will need to use two conversion factors from **Table 2.**)

Table 3 Unit System Equivalents

Type of Measurement	Equivalent
Length	1 in = 2.54 cm 1 yd = 0.91 m 1 mi = 1.61 km
Mass and weight*	1 oz = 28.35 g 1 lb = 0.45 kg 1 ton (short) = 0.91 tonnes (metric tons) 1 lb = 4.45 N
Volume	1 in^3 = 16.39 cm^3 1 qt = 0.95 L 1 gal = 3.78 L
Area	1 in^2 = 6.45 cm^2 1 yd^2 = 0.83 m^2 1 mi^2 = 2.59 km^2 1 acre = 0.40 hectares
Temperature	$°C = \dfrac{(°F - 32)}{1.8}$ $K = °C + 273$

*Weight is measured in standard Earth gravity.

Convert Between Unit Systems **Table 3** gives a list of equivalents that can be used to convert between English and SI units.

Example

If a meterstick has a length of 100 cm, how long is the meterstick in inches?

Step 1   Write the conversion factors for the units given. From **Table 3,** 1 in = 2.54 cm.

$$\frac{1 \text{ in}}{2.54 \text{ cm}} \ and \ \frac{2.54 \text{ cm}}{1 \text{ in}}$$

Step 2   Determine which conversion factor to use. You are converting from cm to in. Use the conversion factor with cm on the bottom.

$$\frac{1 \text{ in}}{2.54 \text{ cm}}$$

Step 3   Multiply the starting quantity and units by the conversion factor. Cancel the starting units with the units in the denominator. Round your answer to the nearest tenth.

$$100 \text{ cm} \times \frac{1 \text{ in}}{2.54 \text{ cm}} = 39.37 \text{ in}$$

The meterstick is about 39.4 in long.

Practice Problem 1 A book has a mass of 5 lb. What is the mass of the book in kg?

Practice Problem 2 Use the equivalent for in and cm (1 in = 2.54 cm) to show how 1 in$^3 \approx$ 16.39 cm^3.

SCIENCE SKILL HANDBOOK

MATH SKILL HANDBOOK

FOLDABLES HANDBOOK

REFERENCE HANDBOOK

GLOSSARY/ GLOSARIO

INDEX

SCIENCE SKILL HANDBOOK

MATH SKILL HANDBOOK

FOLDABLES HANDBOOK

REFERENCE HANDBOOK

GLOSSARY/ GLOSARIO

INDEX

Precision and Significant Digits

When you make a measurement, the value you record depends on the precision of the measuring instrument. This precision is represented by the number of significant digits recorded in the measurement. When counting the number of significant digits, all digits are counted except zeros at the end of a number with no decimal point such as 2,050, and zeros at the beginning of a decimal such as 0.03020. When adding or subtracting numbers with different precision, round the answer to the smallest number of decimal places of any number in the sum or difference. When multiplying or dividing, the answer is rounded to the smallest number of significant digits of any number being multiplied or divided.

Example

The lengths 5.28 and 5.2 are measured in meters. Find the sum of these lengths and record your answer using the correct number of significant digits.

Step 1 Find the sum.

5.28 m	2 digits after the decimal
+ 5.2 m	1 digit after the decimal
10.48 m	

Step 2 Round to one digit after the decimal because the least number of digits after the decimal of the numbers being added is 1.

The sum is 10.5 m.

Practice Problem 1 How many significant digits are in the measurement 7,071,301 m? How many significant digits are in the measurement 0.003010 g?

Practice Problem 2 Multiply 5.28 and 5.2 using the rule for multiplying and dividing. Record the answer using the correct number of significant digits.

Scientific Notation

Many times numbers used in science are very small or very large. Because these numbers are difficult to work with scientists use scientific notation. To write numbers in scientific notation, move the decimal point until only one non-zero digit remains on the left. Then count the number of places you moved the decimal point and use that number as a power of ten. For example, the average distance from the Sun to Mars is 227,800,000,000 m. In scientific notation, this distance is 2.278×10^{11} m. Because you moved the decimal point to the left, the number is a positive power of ten.

The mass of an electron is about 0.000 000 000 000 000 000 000 000 000 000 911 kg. Expressed in scientific notation, this mass is 9.11×10^{-31} kg. Because the decimal point was moved to the right, the number is a negative power of ten.

Example

Earth is 149,600,000 km from the Sun. Express this in scientific notation.

Step 1 Move the decimal point until one non-zero digit remains on the left.

1.496 000 00

Step 2 Count the number of decimal places you have moved. In this case, eight.

Step 2 Show that number as a power of ten, 10^8.

Earth is 1.496×10^8 km from the Sun.

Practice Problem 1 How many significant digits are in 149,600,000 km? How many significant digits are in 1.496×10^8 km?

Practice Problem 2 Parts used in a high performance car must be measured to 7×10^{-6} m. Express this number as a decimal.

Practice Problem 3 A CD is spinning at 539 revolutions per minute. Express this number in scientific notation.

Make and Use Graphs

Data in tables can be displayed in a graph—a visual representation of data. Common graph types include line graphs, bar graphs, and circle graphs.

Line Graph A line graph shows a relationship between two variables that change continuously. The independent variable is changed and is plotted on the x-axis. The dependent variable is observed, and is plotted on the y-axis.

Example

Draw a line graph of the data below from a cyclist in a long-distance race.

Table 4 Bicycle Race Data	
Time (h)	**Distance (km)**
0	0
1	8
2	16
3	24
4	32
5	40

Step 1 Determine the x-axis and y-axis variables. Time varies independently of distance and is plotted on the x-axis. Distance is dependent on time and is plotted on the y-axis.

Step 2 Determine the scale of each axis. The x-axis data ranges from 0 to 5. The y-axis data ranges from 0 to 50.

Step 3 Using graph paper, draw and label the axes. Include units in the labels.

Step 4 Draw a point at the intersection of the time value on the x-axis and corresponding distance value on the y-axis. Connect the points and label the graph with a title, as shown in **Figure 8.**

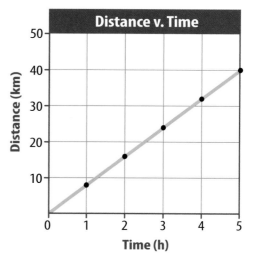

Figure 8 This line graph shows the relationship between distance and time during a bicycle ride.

Practice Problem A puppy's shoulder height is measured during the first year of her life. The following measurements were collected: (3 mo, 52 cm), (6 mo, 72 cm), (9 mo, 83 cm), (12 mo, 86 cm). Graph this data.

Find a Slope The slope of a straight line is the ratio of the vertical change, rise, to the horizontal change, run.

$$\text{Slope} = \frac{\text{vertical change (rise)}}{\text{horizontal change (run)}} = \frac{\text{change in } y}{\text{change in } x}$$

Example

Find the slope of the graph in **Figure 8**.

Step 1 You know that the slope is the change in y divided by the change in x.

$$\text{Slope} = \frac{\text{change in } y}{\text{change in } x}$$

Step 2 Determine the data points you will be using. For a straight line, choose the two sets of points that are the farthest apart.

$$\text{Slope} = \frac{(40 - 0) \text{ km}}{(5 - 0) \text{ h}}$$

Step 3 Find the change in y and x.

$$\text{Slope} = \frac{40 \text{ km}}{5 \text{ h}}$$

Step 4 Divide the change in y by the change in x.

$$\text{Slope} = \frac{8 \text{ km}}{\text{h}}$$

The slope of the graph is 8 km/h.

SCIENCE SKILL HANDBOOK

MATH SKILL HANDBOOK

FOLDABLES HANDBOOK

REFERENCE HANDBOOK

GLOSSARY/ GLOSARIO

INDEX

SCIENCE SKILL HANDBOOK

MATH SKILL HANDBOOK

FOLDABLES HANDBOOK

REFERENCE HANDBOOK

GLOSSARY/ GLOSARIO

INDEX

Bar Graph To compare data that does not change continuously you might choose a bar graph. A bar graph uses bars to show the relationships between variables. The x-axis variable is divided into parts. The parts can be numbers such as years, or a category such as a type of animal. The y-axis is a number and increases continuously along the axis.

Example

A recycling center collects 4.0 kg of aluminum on Monday, 1.0 kg on Wednesday, and 2.0 kg on Friday. Create a bar graph of this data.

Step 1 Select the x-axis and y-axis variables. The measured numbers (the masses of aluminum) should be placed on the y-axis. The variable divided into parts (collection days) is placed on the x-axis.

Step 2 Create a graph grid like you would for a line graph. Include labels and units.

Step 3 For each measured number, draw a vertical bar above the x-axis value up to the y-axis value. For the first data point, draw a vertical bar above Monday up to 4.0 kg.

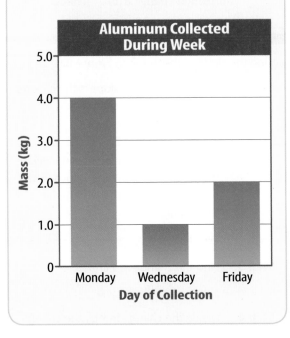

Practice Problem Draw a bar graph of the gases in air: 78% nitrogen, 21% oxygen, 1% other gases.

Circle Graph To display data as parts of a whole, you might use a circle graph. A circle graph is a circle divided into sections that represent the relative size of each piece of data. The entire circle represents 100%, half represents 50%, and so on.

Example

Air is made up of 78% nitrogen, 21% oxygen, and 1% other gases. Display the composition of air in a circle graph.

Step 1 Multiply each percent by 360° and divide by 100 to find the angle of each section in the circle.

$$78\% \times \frac{360°}{100} = 280.8°$$

$$21\% \times \frac{360°}{100} = 75.6°$$

$$1\% \times \frac{360°}{100} = 3.6°$$

Step 2 Use a compass to draw a circle and to mark the center of the circle. Draw a straight line from the center to the edge of the circle.

Step 3 Use a protractor and the angles you calculated to divide the circle into parts. Place the center of the protractor over the center of the circle and line the base of the protractor over the straight line.

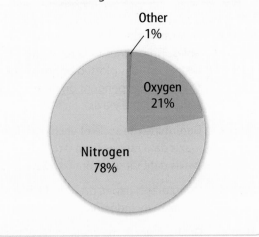

Practice Problem Draw a circle graph to represent the amount of aluminum collected during the week shown in the bar graph to the left.

Student Study Guides & Instructions

By Dinah Zike

1. You will find suggestions for Study Guides, also known as Foldables or books, in each chapter lesson and as a final project. Look at the end of the chapter to determine the project format and glue the Foldables in place as you progress through the chapter lessons.

2. Creating the Foldables or books is simple and easy to do by using copy paper, art paper, and internet printouts. Photocopies of maps, diagrams, or your own illustrations may also be used for some of the Foldables. Notebook paper is the most common source of material for study guides and 83% of all Foldables are created from it. When folded to make books, notebook paper Foldables easily fit into 11″ × 17″ or 12″ × 18″ chapter projects with space left over. Foldables made using photocopy paper are slightly larger and they fit into Projects, but snugly. Use the least amount of glue, tape, and staples needed to assemble the Foldables.

3. Seven of the Foldables can be made using either small or large paper. When 11″ × 17″ or 12″ × 18″ paper is used, these become projects for housing smaller Foldables. Project format boxes are located within the instructions to remind you of this option.

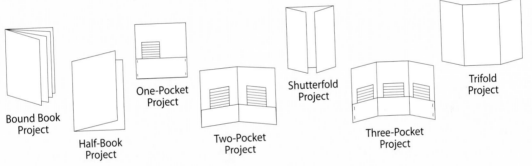

Bound Book Project

Half-Book Project

One-Pocket Project

Two-Pocket Project

Shutterfold Project

Three-Pocket Project

Trifold Project

4. Use one-gallon self-locking plastic bags to store your projects. Place strips of two-inch clear tape along the left, long side of the bag and punch holes through the taped edge. Cut the bottom corners off the bag so it will not hold air. Store this Project Portfolio inside a three-hole binder. To store a large collection of project bags, use a giant laundry-soap box. Holes can be punched in some of the Foldable Projects so they can be stored in a three-hole binder without using a plastic bag. Punch holes in the pocket books before gluing or stapling the pocket.

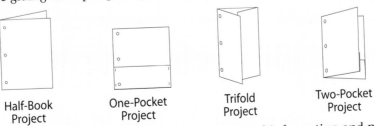

Half-Book Project

One-Pocket Project

Trifold Project

Two-Pocket Project

5. Maximize the use of the projects by collecting additional information and placing it on the back of the project and other unused spaces of the large Foldables.

SCIENCE SKILL HANDBOOK

MATH SKILL HANDBOOK

FOLDABLES HANDBOOK

REFERENCE HANDBOOK

GLOSSARY/ GLOSARIO

INDEX

Half-Book Foldable® By Dinah Zike

Step 1 Fold a sheet of notebook or copy paper in half.

Label the exterior tab and use the inside space to write information.

PROJECT FORMAT
Use 11″ × 17″ or 12″ × 18″ paper on the horizontal axis to make a large project book.

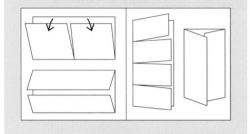

Variations

Paper can be folded horizontally, like a *hamburger* or vertically, like a *hot dog*.

C Half-books can be folded so that one side is ½ inch longer than the other side. A title or question can be written on the extended tab.

- -

Worksheet Foldable or Folded Book® By Dinah Zike

Step 1 Make a half-book (see above) using work sheets, internet print-outs, diagrams, or maps.

Step 2 Fold it in half again.

Variations

A This folded sheet as a small book with two pages can be used for comparing and contrasting, cause and effect, or other skills.

B When the sheet of paper is open, the four sections can be used separately or used collectively to show sequences or steps.

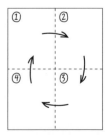

SCIENCE SKILL HANDBOOK

MATH SKILL HANDBOOK

FOLDABLES HANDBOOK

REFERENCE HANDBOOK

GLOSSARY/ GLOSARIO

INDEX

Two-Tab and Concept-Map Foldable® By Dinah Zike

Step 1 Fold a sheet of notebook or copy paper in half vertically or horizontally.

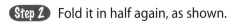

Step 2 Fold it in half again, as shown.

Step 3 Unfold once and cut along the fold line or valley of the top flap to make two flaps.

Variations

A Concept maps can be made by leaving a ½ inch tab at the top when folding the paper in half. Use arrows and labels to relate topics to the primary concept.

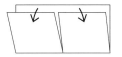

B Use two sheets of paper to make multiple page tab books. Glue or staple books together at the top fold.

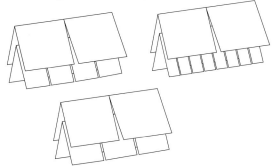

- -

Three-Quarter Foldable® By Dinah Zike

Step 1 Make a two-tab book (see above) and cut the left tab off at the top of the fold line.

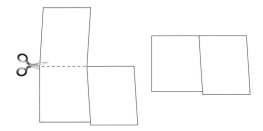

Variations

A Use this book to draw a diagram or a map on the exposed left tab. Write questions about the illustration on the top right tab and provide complete answers on the space under the tab.

B Compose a self-test using multiple choice answers for your questions. Include the correct answer with three wrong responses. The correct answers can be written on the back of the book or upside down on the bottom of the inside page.

SCIENCE SKILL HANDBOOK
MATH SKILL HANDBOOK
FOLDABLES HANDBOOK
REFERENCE HANDBOOK
GLOSSARY/GLOSARIO
INDEX

SCIENCE SKILL HANDBOOK

MATH SKILL HANDBOOK

FOLDABLES HANDBOOK

REFERENCE HANDBOOK

GLOSSARY/ GLOSARIO

INDEX

Three-Tab Foldable® By Dinah Zike

Step 1 Fold a sheet of paper in half horizontally.

Step 2 Fold into thirds.

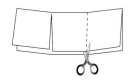

Step 3 Unfold and cut along the folds of the top flap to make three sections.

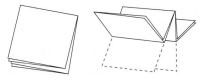

Variations

A Before cutting the three tabs draw a Venn diagram across the front of the book.

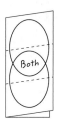

B Make a space to use for titles or concept maps by leaving a ½ inch tab at the top when folding the paper in half.

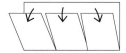

Four-Tab Foldable® By Dinah Zike

Step 1 Fold a sheet of paper in half horizontally.

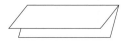

Step 2 Fold in half and then fold each half as shown below.

Step 3 Unfold and cut along the fold lines of the top flap to make four tabs.

Variations

A Make a space to use for titles or concept maps by leaving a ½ inch tab at the top when folding the paper in half.

B Use the book on the vertical axis, with or without an extended tab.

Folding Fifths for a Foldable® By Dinah Zike

Step 1 Fold a sheet of paper in half horizontally.

Step 2 Fold again so one-third of the paper is exposed and two-thirds are covered.

Step 3 Fold the two-thirds section in half.

Step 4 Fold the one-third section, a single thickness, backward to make a fold line.

Variations

A Unfold and cut along the fold lines to make five tabs.

B Make a five-tab book with a ½ inch tab at the top (see two-tab instructions).

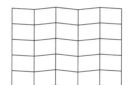

C Use 11" × 17" or 12" × 18" paper and fold into fifths for a five-column and/or row table or chart.

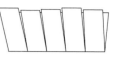

- -

Folded Table or Chart, and Trifold Foldable® By Dinah Zike

Step 1 Fold a sheet of paper in the required number of vertical columns for the table or chart.

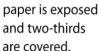

Step 2 Fold the horizontal rows needed for the table or chart.

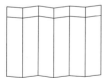

Variations

A Make a trifold by folding the paper into thirds vertically or horizontally.

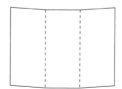

PROJECT FORMAT
Use 11" × 17" or 12" × 18" paper and fold it to make a large trifold project book or larger tables and charts.

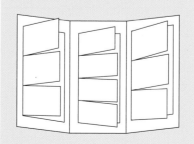

B Make a trifold book. Unfold it and draw a Venn diagram on the inside.

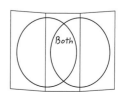

SCIENCE SKILL HANDBOOK

MATH SKILL HANDBOOK

FOLDABLES HANDBOOK

REFERENCE HANDBOOK

GLOSSARY/ GLOSARIO

INDEX

SCIENCE SKILL HANDBOOK

MATH SKILL HANDBOOK

FOLDABLES HANDBOOK

REFERENCE HANDBOOK

GLOSSARY/ GLOSARIO

INDEX

Two or Three-Pockets Foldable® By Dinah Zike

Step 1 Fold up the long side of a horizontal sheet of paper about 5 cm.

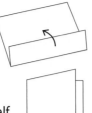

Step 2 Fold the paper in half.

Step 3 Open the paper and glue or staple the outer edges to make two compartments.

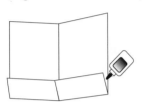

Variations

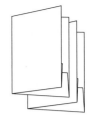

A Make a multi-page booklet by gluing several pocket books together.

B Make a three-pocket book by using a trifold (see previous instructions).

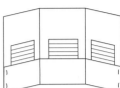

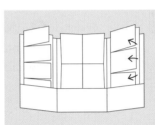

PROJECT FORMAT
Use 11″ × 17″ or 12″ × 18″ paper and fold it horizontally to make a large multi-pocket project.

- -

Matchbook Foldable® By Dinah Zike

Step 1 Fold a sheet of paper almost in half and make the back edge about 1–2 cm longer than the front edge.

Step 2 Find the midpoint of the shorter flap.

Step 3 Open the paper and cut the short side along the midpoint making two tabs.

Step 4 Close the book and fold the tab over the short side.

Variations

A Make a single-tab matchbook by skipping Steps 2 and 3.

B Make two smaller matchbooks by cutting the single-tab matchbook in half.

Shutterfold Foldable® By Dinah Zike

Step 1 Begin as if you were folding a vertical sheet of paper in half, but instead of creasing the paper, pinch it to show the midpoint.

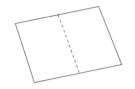

PROJECT FORMAT
Use 11" × 17" or 12" × 18" paper and fold it to make a large shutterfold project.

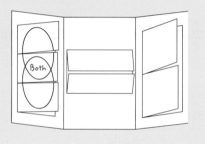

Step 2 Fold the top and bottom to the middle and crease the folds.

Variations

A Use the shutterfold on the horizontal axis.

B Create a center tab by leaving .5–2 cm between the flaps in Step 2.

Four-Door Foldable® By Dinah Zike

Step 1 Make a shutterfold (see above).

Step 2 Fold the sheet of paper in half.

Step 3 Open the last fold and cut along the inside fold lines to make four tabs.

Variations

A Use the four-door book on the opposite axis.

B Create a center tab by leaving .5–2 cm between the flaps in Step 1.

SCIENCE SKILL HANDBOOK

MATH SKILL HANDBOOK

FOLDABLES HANDBOOK

REFERENCE HANDBOOK

GLOSSARY/ GLOSARIO

INDEX

Science Skill Handbook

Math Skill Handbook

Foldables Handbook

Reference Handbook

Glossary/Glosario

Index

Bound Book Foldable® By Dinah Zike

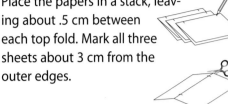

Step 1 Fold three sheets of paper in half. Place the papers in a stack, leaving about .5 cm between each top fold. Mark all three sheets about 3 cm from the outer edges.

Step 2 Using two of the sheets, cut from the outer edges to the marked spots on each side. On the other sheet, cut between the marked spots.

Step 3 Take the two sheets from Step 1 and slide them through the cut in the third sheet to make a 12-page book.

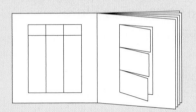

Step 4 Fold the bound pages in half to form a book.

Variation

A Use two sheets of paper to make an eight-page book, or increase the number of pages by using more than three sheets.

PROJECT FORMAT
Use two or more sheets of 11″ × 17″ or 12″ × 18″ paper and fold it to make a large bound book project.

Accordian Foldable® By Dinah Zike

Step 1 Fold the selected paper in half vertically, like a *hamburger*.

Step 2 Cut each sheet of folded paper in half along the fold lines.

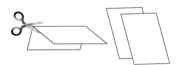

Step 3 Fold each half-sheet almost in half, leaving a 2 cm tab at the top.

Step 4 Fold the top tab over the short side, then fold it in the opposite direction.

Variations

A Glue the straight edge of one paper inside the tab of another sheet. Leave a tab at the end of the book to add more pages.

B Tape the straight edge of one paper to the tab of another sheet, or just tape the straight edges of nonfolded paper end to end to make an accordian.

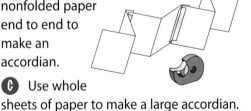

C Use whole sheets of paper to make a large accordian.

Layered Foldable® By Dinah Zike

Step 1 Stack two sheets of paper about 1–2 cm apart. Keep the right and left edges even.

Step 2 Fold up the bottom edges to form four tabs. Crease the fold to hold the tabs in place.

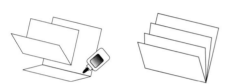

Step 3 Staple along the folded edge, or open and glue the papers together at the fold line.

Variations

A Rotate the book so the fold is at the top or to the side.

B Extend the book by using more than two sheets of paper.

Envelope Foldable® By Dinah Zike

Step 1 Fold a sheet of paper into a *taco*. Cut off the tab at the top.

Step 2 Open the *taco* and fold it the opposite way making another *taco* and an X-fold pattern on the sheet of paper.

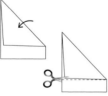

Step 3 Cut a map, illustration, or diagram to fit the inside of the envelope.

Step 4 Use the outside tabs for labels and inside tabs for writing information.

Variations

A Use 11″ × 17″ or 12″ × 18″ paper to make a large envelope.

B Cut off the points of the four tabs to make a window in the middle of the book.

SCIENCE SKILL HANDBOOK

MATH SKILL HANDBOOK

FOLDABLES HANDBOOK

REFERENCE HANDBOOK

GLOSSARY/ GLOSARIO

INDEX

Sentence Strip Foldable® By Dinah Zike

Step 1 Fold two sheets of paper in half vertically, like a *hamburger*.

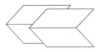

Step 2 Unfold and cut along fold lines making four half sheets.

Step 3 Fold each half sheet in half horizontally, like a *hot dog*.

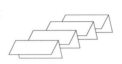

Step 4 Stack folded horizontal sheets evenly and staple together on the left side.

Step 5 Open the top flap of the first sentence strip and make a cut about 2 cm from the stapled edge to the fold line. This forms a flap that can be raisied and lowered. Repeat this step for each sentence strip.

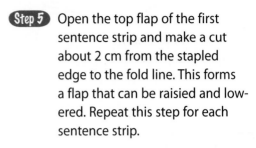

Variations

A Expand this book by using more than two sheets of paper.

B Use whole sheets of paper to make large books.

Pyramid Foldable® By Dinah Zike

Step 1 Fold a sheet of paper into a *taco*. Crease the fold line, but do not cut it off.

Step 2 Open the folded sheet and refold it like a *taco* in the opposite direction to create an X-fold pattern.

Step 3 Cut one fold line as shown, stopping at the center of the X-fold to make a flap.

Step 4 Outline the fold lines of the X-fold. Label the three front sections and use the inside spaces for notes. Use the tab for the title.

Step 5 Glue the tab into a project book or notebook. Use the space under the pyramid for other information.

Step 6 To display the pyramid, fold the flap under and secure with a paper clip, if needed.

SCIENCE SKILL HANDBOOK

MATH SKILL HANDBOOK

FOLDABLES HANDBOOK

REFERENCE HANDBOOK

GLOSSARY/ GLOSARIO

INDEX

Single-Pocket or One-Pocket Foldable® By Dinah Zike

Step 1 Using a large piece of paper on a vertical axis, fold the bottom edge of the paper upwards, about 5 cm.

Step 2 Glue or staple the outer edges to make a large pocket.

PROJECT FORMAT
Use 11" × 17" or 12" × 18" paper and fold it vertically or horizontally to make a large pocket project.

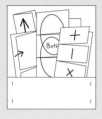

Variations

A Make the one-pocket project using the paper on the horizontal axis.

B To store materials securely inside, fold the top of the paper almost to the center, leaving about 2–4 cm between the paper edges. Slip the Foldables through the opening and under the top and bottom pockets.

Multi-Tab Foldable® By Dinah Zike

Step 1 Fold a sheet of notebook paper in half like a *hot dog*.

Step 2 Open the paper and on one side cut every third line. This makes ten tabs on wide ruled notebook paper and twelve tabs on college ruled.

Step 3 Label the tabs on the front side and use the inside space for definitions or other information.

Variation

A Make a tab for a title by folding the paper so the holes remain uncovered. This allows the notebook Foldable to be stored in a three-hole binder.

SCIENCE SKILL HANDBOOK

MATH SKILL HANDBOOK

FOLDABLES HANDBOOK

REFERENCE HANDBOOK

GLOSSARY/ GLOSARIO

INDEX

PERIODIC TABLE OF THE ELEMENTS

Element — Hydrogen
Atomic number — 1
Symbol — H
Atomic mass — 1.01
State of matter

- Gas
- Liquid
- Solid
- Synthetic

A column in the periodic table is called a **group.**

A row in the periodic table is called a **period.**

	1	2	3	4	5	6	7	8	9
1	Hydrogen 1 H 1.01								
2	Lithium 3 Li 6.94	Beryllium 4 Be 9.01							
3	Sodium 11 Na 22.99	Magnesium 12 Mg 24.31							
4	Potassium 19 K 39.10	Calcium 20 Ca 40.08	Scandium 21 Sc 44.96	Titanium 22 Ti 47.87	Vanadium 23 V 50.94	Chromium 24 Cr 52.00	Manganese 25 Mn 54.94	Iron 26 Fe 55.85	Cobalt 27 Co 58.93
5	Rubidium 37 Rb 85.47	Strontium 38 Sr 87.62	Yttrium 39 Y 88.91	Zirconium 40 Zr 91.22	Niobium 41 Nb 92.91	Molybdenum 42 Mo 95.96	Technetium 43 Tc (98)	Ruthenium 44 Ru 101.07	Rhodium 45 Rh 102.91
6	Cesium 55 Cs 132.91	Barium 56 Ba 137.33	Lanthanum 57 La 138.91	Hafnium 72 Hf 178.49	Tantalum 73 Ta 180.95	Tungsten 74 W 183.84	Rhenium 75 Re 186.21	Osmium 76 Os 190.23	Iridium 77 Ir 192.22
7	Francium 87 Fr (223)	Radium 88 Ra (226)	Actinium 89 Ac (227)	Rutherfordium 104 Rf (267)	Dubnium 105 Db (268)	Seaborgium 106 Sg (271)	Bohrium 107 Bh (272)	Hassium 108 Hs (270)	Meitnerium 109 Mt (276)

The number in parentheses is the mass number of the longest lived isotope for that element.

Lanthanide series	Cerium 58 Ce 140.12	Praseodymium 59 Pr 140.91	Neodymium 60 Nd 144.24	Promethium 61 Pm (145)	Samarium 62 Sm 150.36	Europium 63 Eu 151.96
Actinide series	Thorium 90 Th 232.04	Protactinium 91 Pa 231.04	Uranium 92 U 238.03	Neptunium 93 Np (237)	Plutonium 94 Pu (244)	Americium 95 Am (243)

Metal

Metalloid

Nonmetal

Recently discovered

		13	14	15	16	17	18
							Helium 2 He 4.00

| 10 | 11 | 12 | Boron 5 B 10.81 | Carbon 6 C 12.01 | Nitrogen 7 N 14.01 | Oxygen 8 O 16.00 | Fluorine 9 F 19.00 | Neon 10 Ne 20.18 |

Aluminum 13 Al 26.98 | Silicon 14 Si 28.09 | Phosphorus 15 P 30.97 | Sulfur 16 S 32.07 | Chlorine 17 Cl 35.45 | Argon 18 Ar 39.95

Nickel 28 Ni 58.69	Copper 29 Cu 63.55	Zinc 30 Zn 65.38	Gallium 31 Ga 69.72	Germanium 32 Ge 72.64	Arsenic 33 As 74.92	Selenium 34 Se 78.96	Bromine 35 Br 79.90	Krypton 36 Kr 83.80
Palladium 46 Pd 106.42	Silver 47 Ag 107.87	Cadmium 48 Cd 112.41	Indium 49 In 114.82	Tin 50 Sn 118.71	Antimony 51 Sb 121.76	Tellurium 52 Te 127.60	Iodine 53 I 126.90	Xenon 54 Xe 131.29
Platinum 78 Pt 195.08	Gold 79 Au 196.97	Mercury 80 Hg 200.59	Thallium 81 Tl 204.38	Lead 82 Pb 207.20	Bismuth 83 Bi 208.98	Polonium 84 Po (209)	Astatine 85 At (210)	Radon 86 Rn (222)
Darmstadtium 110 Ds (281)	Roentgenium 111 Rg (280)	Copernicium 112 Cn (285)	* Ununtrium 113 Uut (284)	* Ununquadium 114 Uuq (289)	* Ununpentium 115 Uup (288)	* Ununhexium 116 Uuh (293)		* Ununoctium 118 Uuo (294)

* The names and symbols for elements 113–116 and 118 are temporary. Final names will be selected when the elements' discoveries are verified.

| Gadolinium 64 Gd 157.25 | Terbium 65 Tb 158.93 | Dysprosium 66 Dy 162.50 | Holmium 67 Ho 164.93 | Erbium 68 Er 167.26 | Thulium 69 Tm 168.93 | Ytterbium 70 Yb 173.05 | Lutetium 71 Lu 174.97 |
| Curium 96 Cm (247) | Berkelium 97 Bk (247) | Californium 98 Cf (251) | Einsteinium 99 Es (252) | Fermium 100 Fm (257) | Mendelevium 101 Md (258) | Nobelium 102 No (259) | Lawrencium 103 Lr (262) |

SCIENCE SKILL HANDBOOK

MATH SKILL HANDBOOK

FOLDABLES HANDBOOK

REFERENCE HANDBOOK

GLOSSARY/ GLOSARIO

INDEX

SCIENCE SKILL HANDBOOK

MATH SKILL HANDBOOK

FOLDABLES HANDBOOK

REFERENCE HANDBOOK

GLOSSARY/ GLOSARIO

INDEX

Diversity of Life: Classification of Living Organisms

A six-kingdom system of classification of organisms is used today. Two kingdoms—Kingdom Archaebacteria and Kingdom Eubacteria—contain organisms that do not have a nucleus and that lack membrane-bound structures in the cytoplasm of their cells. The members of the other four kingdoms have a cell or cells that contain a nucleus and structures in the cytoplasm, some of which are surrounded by membranes. These kingdoms are Kingdom Protista, Kingdom Fungi, Kingdom Plantae, and Kingdom Animalia.

Kingdom Archaebacteria

one-celled; some absorb food from their surroundings; some are photosynthetic; some are chemosynthetic; many are found in extremely harsh environments including salt ponds, hot springs, swamps, and deep-sea hydrothermal vents

Kingdom Eubacteria

one-celled; most absorb food from their surroundings; some are photosynthetic; some are chemosynthetic; many are parasites; many are round, spiral, or rod-shaped; some form colonies

Kingdom Protista

Phylum Euglenophyta one-celled; photosynthetic or take in food; most have one flagellum; euglenoids

Kingdom Eubacteria
Bacillus anthracis

Phylum Chlorophyta
Desmids

Phylum Bacillariophyta one-celled; photosynthetic; have unique double shells made of silica; diatoms

Phylum Dinoflagellata one-celled; photosynthetic; contain red pigments; have two flagella; dinoflagellates

Phylum Chlorophyta one-celled, many-celled, or colonies; photosynthetic; contain chlorophyll; live on land, in freshwater, or salt water; green algae

Phylum Rhodophyta most are many-celled; photosynthetic; contain red pigments; most live in deep, saltwater environments; red algae

Phylum Phaeophyta most are many-celled; photosynthetic; contain brown pigments; most live in saltwater environments; brown algae

Phylum Rhizopoda one-celled; take in food; are free-living or parasitic; move by means of pseudopods; amoebas

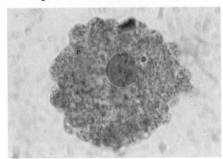

Amoeba

Phylum Zoomastigina one-celled; take in food; free-living or parasitic; have one or more flagella; zoomastigotes

Phylum Ciliophora one-celled; take in food; have large numbers of cilia; ciliates

Phylum Sporozoa one-celled; take in food; have no means of movement; are parasites in animals; sporozoans

Phylum Myxomycota
Slime mold

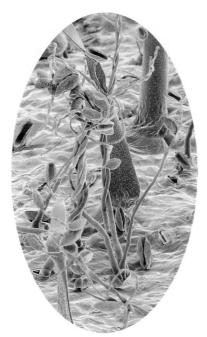

Phylum Oomycota
Phytophthora infestans

Phyla Myxomycota and Acrasiomycota one- or many-celled; absorb food; change form during life cycle; cellular and plasmodial slime molds

Phylum Oomycota many-celled; are either parasites or decomposers; live in freshwater or salt water; water molds, rusts and downy mildews

Kingdom Fungi

Phylum Zygomycota many-celled; absorb food; spores are produced in sporangia; zygote fungi; bread mold

Phylum Ascomycota one- and many-celled; absorb food; spores produced in asci; sac fungi; yeast

Phylum Basidiomycota many-celled; absorb food; spores produced in basidia; club fungi; mushrooms

Phylum Deuteromycota members with unknown reproductive structures; imperfect fungi; *Penicillium*

Phylum Mycophycota organisms formed by symbiotic relationship between an ascomycote or a basidiomycote and green alga or cyanobacterium; lichens

Lichens

SCIENCE SKILL HANDBOOK

MATH SKILL HANDBOOK

FOLDABLES HANDBOOK

REFERENCE HANDBOOK

GLOSSARY/ GLOSARIO

INDEX

SCIENCE SKILL HANDBOOK

MATH SKILL HANDBOOK

FOLDABLES HANDBOOK

REFERENCE HANDBOOK

GLOSSARY/ GLOSARIO

INDEX

Kingdom Plantae

Divisions Bryophyta (mosses), **Anthocerophyta** (hornworts), **Hepaticophyta** (liverworts), **Psilophyta** (whisk ferns) many-celled non-vascular plants; reproduce by spores produced in capsules; green; grow in moist, land environments

Division Lycophyta many-celled vascular plants; spores are produced in conelike structures; live on land; are photosynthetic; club mosses

Division Arthrophyta vascular plants; ribbed and jointed stems; scalelike leaves; spores produced in conelike structures; horsetails

Division Pterophyta vascular plants; leaves called fronds; spores produced in clusters of sporangia called sori; live on land or in water; ferns

Division Ginkgophyta deciduous trees; only one living species; have fan-shaped leaves with branching veins and fleshy cones with seeds; ginkgoes

Division Cycadophyta palmlike plants; have large, featherlike leaves; produces seeds in cones; cycads

Division Coniferophyta deciduous or evergreen; trees or shrubs; have needlelike or scalelike leaves; seeds produced in cones; conifers

Division Anthophyta
Tomato plant

Phylum Platyhelminthes
Flatworm

Division Gnetophyta shrubs or woody vines; seeds are produced in cones; division contains only three genera; gnetum

Division Anthophyta dominant group of plants; flowering plants; have fruits with seeds

Kingdom Animalia

Phylum Porifera aquatic organisms that lack true tissues and organs; are asymmetrical and sessile; sponges

Phylum Cnidaria radially symmetrical organisms; have a digestive cavity with one opening; most have tentacles armed with stinging cells; live in aquatic environments singly or in colonies; includes jellyfish, corals, hydra, and sea anemones

Phylum Platyhelminthes bilaterally symmetrical worms; have flattened bodies; digestive system has one opening; parasitic and free-living species; flatworms

Division Bryophyta
Liverwort

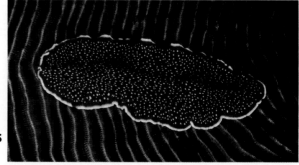

Phylum Chordata

Phylum Nematoda round, bilaterally symmetrical body; have digestive system with two openings; free-living forms and parasitic forms; roundworms

Phylum Mollusca soft-bodied animals, many with a hard shell and soft foot or footlike appendage; a mantle covers the soft body; aquatic and terrestrial species; includes clams, snails, squid, and octopuses

Phylum Annelida bilaterally symmetrical worms; have round, segmented bodies; terrestrial and aquatic species; includes earthworms, leeches, and marine polychaetes

Phylum Arthropoda largest animal group; have hard exoskeletons, segmented bodies, and pairs of jointed appendages; land and aquatic species; includes insects, crustaceans, and spiders

Phylum Echinodermata marine organisms; have spiny or leathery skin and a water-vascular system with tube feet; are radially symmetrical; includes sea stars, sand dollars, and sea urchins

Phylum Chordata organisms with internal skeletons and specialized body systems; most have paired appendages; all at some time have a notochord, nerve cord, gill slits, and a post-anal tail; include fish, amphibians, reptiles, birds, and mammals

SCIENCE SKILL HANDBOOK

MATH SKILL HANDBOOK

FOLDABLES HANDBOOK

REFERENCE HANDBOOK

GLOSSARY/ GLOSARIO

INDEX

Use and Care of a Microscope

SCIENCE SKILL HANDBOOK

MATH SKILL HANDBOOK

FOLDABLES HANDBOOK

REFERENCE HANDBOOK

GLOSSARY/ GLOSARIO

INDEX

Eyepiece Contains magnifying lenses you look through.

Arm Supports the body tube.

Low-power objective Contains the lens with the lowest power magnification.

Stage clips Hold the microscope slide in place.

Coarse adjustment Focuses the image under low power.

Fine adjustment Sharpens the image under high magnification.

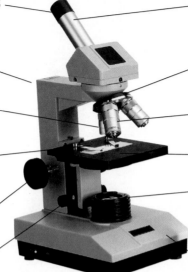

Body tube Connects the eyepiece to the revolving nosepiece.

Revolving nosepiece Holds and turns the objectives into viewing position.

High-power objective Contains the lens with the highest magnification.

Stage Supports the microscope slide.

Light source Provides light that passes upward through the diaphragm, the specimen, and the lenses.

Base Provides support for the microscope.

Caring for a Microscope

1. Always carry the microscope holding the arm with one hand and supporting the base with the other hand.
2. Don't touch the lenses with your fingers.
3. The coarse adjustment knob is used only when looking through the lowest-power objective lens. The fine adjustment knob is used when the high-power objective is in place.
4. Cover the microscope when you store it.

Using a Microscope

1. Place the microscope on a flat surface that is clear of objects. The arm should be toward you.
2. Look through the eyepiece. Adjust the diaphragm so light comes through the opening in the stage.
3. Place a slide on the stage so the specimen is in the field of view. Hold it firmly in place by using the stage clips.

4. Always focus with the coarse adjustment and the low-power objective lens first. After the object is in focus on low power, turn the nosepiece until the high-power objective is in place. Use ONLY the fine adjustment to focus with the high-power objective lens.

Making a Wet-Mount Slide

1. Carefully place the item you want to look at in the center of a clean, glass slide. Make sure the sample is thin enough for light to pass through.
2. Use a dropper to place one or two drops of water on the sample.
3. Hold a clean coverslip by the edges and place it at one edge of the water. Slowly lower the coverslip onto the water until it lies flat.
4. If you have too much water or a lot of air bubbles, touch the edge of a paper towel to the edge of the coverslip to draw off extra water and draw out unwanted air.

Glossary/Glosario

Cómo usar el glosario en español:
1. Busca el término en inglés que desees encontrar.
2. El término en español, junto con la definición, se encuentran en la columna de la derecha.

Pronunciation Key

Use the following key to help you sound out words in the glossary.

a . . . . ba**ck** (BAK)		**ew**. . . . . f**oo**d (FEWD)	
ay . . . . d**ay** (DAY)		**yoo** . . . . p**u**re (PYOOR)	
ah . . . . f**a**ther (FAH thur)		**yew**. . . . . f**ew** (FYEW)	
ow . . . . fl**ow**er (FLOW ur)		**uh**. . . . . comm**a** (CAH muh)	
ar . . . . **car** (CAR)		**u (+ con)** . . . . r**u**b (RUB)	
e . . . . l**e**ss (LES)		**sh** . . . . **sh**elf (SHELF)	
ee . . . . l**ea**f (LEEF)		**ch** . . . . na**t**ure (NAY chur)	
ih . . . . tr**i**p (TRIHP)		**g** . . . . **g**ift (GIHFT)	
i (i + com + e)	**i**dea (i DEE uh)	**j** . . . . **g**em (JEM)	
oh . . . . g**o** (GOH)		**ing** . . . . s**ing** (SING)	
aw . . . . s**o**ft (SAWFT)		**zh** . . . . vi**si**on (VIH zhun)	
or . . . . **or**bit (OR buht)		**k**. . . . . ca**k**e (KAYK)	
oy . . . . c**oi**n (COYN)		**s** . . . . **s**eed, **c**ent (SEED, SENT)	
oo . . . . f**oo**t (FOOT)		**z** . . . . **z**one, rai**s**e (ZOHN, RAYZ)	

English	**A**	**Español**

alga (plural, algae)/antibody

alga (plural, algas)/anticuerpo

alga (plural, algae): a plantlike protist that produces food through photosynthesis using light energy and carbon dioxide. (p. 266)

alternation of generations: process that occurs when the life cycle of an organism alternates between diploid and haploid generations. (p. 352)

amoeba (uh MEE buh): one common sarcodine with an unusual adaptation for movement and getting nutrients. (p. 271)

antibiotic (an ti bi AH tihk): a medicine that stops the growth and reproduction of bacteria. (p. 242)

antibody: a protein that can attach to a pathogen, making the pathogen useless. (p. 251)

alga (plural, algas): protista parecida a una planta que produce el alimento por medio de la fotosíntesis, usando la energía lumínica y el dióxido de carbono. (pág. 266)

alternancia de generaciones: proceso que ocurre cuando el ciclo de vida de un organismo se alterna entre generaciones diploides y haploides. (pág. 352)

ameba: sarcodina común con una adaptación inusual para moverse y obtener nutrientes. (pág. 271)

antibiótico: medicina que detiene el crecimiento y reproducción de las bacterias. (pág. 242)

anticuerpo: proteína que se adhiere a un patógeno y lo hace inútil. (pág. 251)

ascus (AS kuhs): the reproductive structure where spores develop on sac fungi. (p. 279)

ascus: estructura reproductiva donde se desarrollan las esporas en un hongo con saco. (pág. 279)

B

bacterium: a microscopic prokaryote. (p. 231)

basidium (buh SIH dee uhm): reproductive structure that produces sexual spores inside the basidiocarp. (p. 278)

bioremediation (bi oh rih mee dee AY shun): the use of organisms, such as bacteria, to clean up environmental pollution. (p. 241)

bacteria: procariota microscópica. (pág. 231)

basidio: estructura reproductiva que produce esporas sexuales en el interior de un basidiocarpo. (pág. 278)

biorremediación: uso de microorganismos, como bacterias, para limpiar la contaminación del medioambiente. (pág. 241)

C

cambium: a layer of tissue that produces new vascular tissue and grows between xylem and phloem. (p. 314)

cellular respiration: a series of chemical reactions that convert the energy in food molecules into a usable form of energy called ATP. (p. 336)

cellulose: an organic compound made of chains of glucose molecules. (p. 299)

cilia (SIH lee uh): short, hairlike structures that grow on the surface of some protists. (p. 270)

conjugation (kahn juh GAY shun): a process during which two bacteria of the same species attach to each other and combine their genetic material. (p. 234)

cuticle: a waxy, protective layer on the leaves, stems, and flowers of plants. (p. 299)

cámbium: capa de tejido que produce tejido vascular nuevo y crece en medio del xilema y el floema. (pág. 314)

respiración celular: serie de reacciones químicas que convierten la energía de las moléculas de alimento en una forma de energía utilizable llamada ATP. (pág. 336)

celulosa: compuesto orgánico constituido de cadenas de moléculas de glucosa. (pág. 299)

cilios: estructuras cortas parecidas a un cabello que crecen en la superficie de algunos protistas. (pág. 270)

conjugación: proceso durante el cual dos bacterias de la misma especie se adhieren una a la otra y combinan sus material genético. (pág. 234)

cutícula: capa cerosa de protección que tienen las hojas, los tallos y las flores de las plantas. (pág. 299)

D

decomposition: the breaking down of dead organisms and organic waste. (p. 240)

diatom (DI uh tahm): a type of microscopic plantlike protist with a hard outer wall. (p. 267)

descomposición: degradación de organismos muertos y desecho orgánico. (pág. 240)

diatomea: tipo de protista microscópico parecido a una planta que tiene una pared externa dura. (pág. 267)

E

embryo: an immature diploid plant that develops from the zygote. (p. 354)

embrión: planta diploide inmadura que se desarrolla de un zigoto. (pág. 354)

endospore (EN doh spor): a thick internal wall that a bacterium builds around its chromosome and part of its cytoplasm. (p. 235)

endospora: pared interna gruesa que una bacteria produce alrededor del cromosoma y parte del citoplasma. (pág. 235)

fission: cell division that forms two genetically identical cells. (p. 234)

flagellum (fluh JEH lum): a long, whiplike structure on many bacteria. (p. 234)

frond: a leaf of a fern. (p. 309)

fruit: plant structure that contains one or more seeds; develops from the ovary and sometimes other parts of the flower. (p. 357)

fisión: división celular que forma dos células genéticamente idénticas. (pág. 234)

flagelo: estructura larga similar a un látigo que tienen muchas bacterias. (pág. 234)

fronda: hoja de un helecho. (pág. 309)

fruta: estructura de la planta que contiene una o más semillas; se desarrolla del ovario y algunas veces de otras partes de la flor. (pág. 357)

hyphae (HI fee): long, threadlike structures that make up the body of fungi and also form an underground structure that absorbs minerals and water. (p. 277)

hifas: estructuras largas en forma de filamentos que constituyen el cuerpo de los hongos y que también forman una estructura subterránea que absorbe minerales y agua. (pág. 277)

lichen (LI kun): a structure formed when fungi and certain other photosynthetic organisms grow together. (p. 284)

líquen: estructura formada cuando crecen juntos los hongos y algunos organismos que realizan la fotosíntesis. (pág. 284)

mycelium (mi SEE lee um): an underground network of hyphae. (p. 277)

mychorriza (mi kuh RI zuh): a structure formed when the roots of a plant and the hyphae of a fungus weave together. (p. 282)

micelio: red subterránea de hifas. (pág. 277)

micorriza: estructura formada cuando las raíces de una planta y las hifas de de un hongo se entrelazan. (pág. 282)

N

nitrogen fixation (NI truh jun • fihk SAY shun): the process that changes atmospheric nitrogen into nitrogen compounds that are usable by living things. (p. 240)

fijación de nitrógeno: proceso por el cual el nitrógeno atmosférico se transforma en compuestos de nitrógeno que los seres vivos usan. (pág. 240)

O

ovary: structure located at the base of the style of a flower that contains one or more ovules. (p. 356)

ovario: estructura situado en la base del estilo de una flor que contiene uno o más óvulos. (pág. 356)

ovule: female reproductive structure of a seed plant where the haploid egg develops. (p. 354)

óvulo: estructura reproductiva femenina de la semilla de una planta donde el huevo haploide se desarrolla. (pág. 354)

P

paramecium (pa ruh MEE see um): a protist with cilia and two types of nuclei. (p. 270)

pasteurization (pas chuh ruh ZAY shun): a process of heating food or liquid to a temperature that kills most harmful bacteria. (p. 243)

pathogen (PA thuh jun): an agent that causes disease. (p. 242)

phloem (FLOH em): a type of vascular tissue that carries dissolved sugars throughout a plant. (p. 315)

photoperiodism: a plant's response to the number of hours of darkness in its environment. (p. 344)

photosynthesis (foh toh SIHN thuh sus): a series of chemical reactions that convert light energy, water, and carbon dioxide into the food-energy molecule glucose and give off oxygen. (p. 334)

pistil: female reproductive organ of a flower. (p. 356)

plant hormone: a substance that acts as a chemical messenger within a plant. (p. 345)

pollen (PAH lun) grain: spore that forms from tissue in a male reproductive structure of a seed plant. (p. 354)

pollination (pah luh NAY shun): the process that occurs when pollen grains land on a female reproductive structure of a plant that is the same species as the pollen grains. (p. 354)

producer: an organism that uses an outside energy source, such as the Sun, and produces its own food. (p. 298)

protist: a member of a group of eukaryotic organisms, which have a membrane-bound nucleus. (p. 265)

protozoan (proh tuh ZOH un): a protist that resembles a tiny animal. (p. 270)

pseudopod: a temporary "foot" that forms as an organism pushes part of its body outward. (p. 271)

paramecio: protista con cilios y dos tipos de núcleos. (pág. 270)

pasteurización: proceso en el cual se calientan los alimentos o líquidos para matar la mayoría de bacterias dañinas. (pág. 243)

patógeno: agente que causa enfermedad. (pág. 242)

floema: tipo de tejido vascular que transporta azúcares disueltos por toda la planta. (pág. 315)

fotoperiodismo: respuesta de una planta al número de horas de oscuridad en su medioambiente. (pág. 344)

fotosíntesis: serie de reacciones químicas que convierte la energía lumínica, el agua y el dióxido de carbono en glucosa, una molécula de energía alimentaria, y libera oxígeno. (pág. 334)

pistilo: órgano reproductor femenino de una flor. (pág. 356)

fitohormona: sustancia que actúa como mensajero químico dentro de una planta. (pág. 345)

grano de polen: espora que se forma de tejido en una estructura reproductiva masculina de una planta de semilla. (pág. 354)

polinización: proceso que ocurre cuando los granos de polen posan sobre una estructura reproductiva femenina de una planta que es de la misma especie que los granos de polen. (pág. 354)

productor: organismo que usa una fuente de energía externa, como el Sol, y fabricar su propio alimento. (pág. 298)

protista: miembro de un grupo de organismos eucarióticos que tienen un núcleo limitado por una membrana. (pág. 265)

protozoario: protista que parece un animal pequeño. (pág. 270)

seudópodo: "pata" temporal que se forma a medida que el organismo empuja parte del cuerpo hacia afuera. (pág. 271)

R

rhizoid: a structure that anchors a nonvascular plant to a surface. (p. 307)

rizoide: estructura que sujeta una planta no vascular a una superficie. (pág. 307)

S

seed: a plant embryo, its food supply, and a protective covering. (p. 354)

spore: a daughter cell produced from a haploid structure. (p. 352)

stamen: the male reproductive organ of a flower. (p. 356)

stimulus (STIHM yuh lus): a change in an organism's environment that causes a response. (p. 341)

stoma (STOH muh): a small opening in the epidermis, or surface layer, of a leaf. (p. 317)

semilla: embrión de una planta, su suministro de alimento y cubierta protectora. (pág. 354)

espora: célula hija producida de una estructura haploide. (pág. 352)

estambre: órgano reproductor masculino de una flor. (pág. 356)

estímulo: un cambio en el medioambiente de un organismo que causa una respuesta. (pág. 341)

estoma: abertura pequeña en la epidermis, o capa superficial, de una hoja. (pág. 317)

T

tropism (TROH pih zum): plant growth toward or away from an external stimulus. (p. 342)

tropismo: crecimiento de las plantas hacia o lejos de un estímulo externo. (pág. 342)

V

vaccine: a mixture containing material from one or more deactivated pathogens, such as viruses. (p. 252)

vascular tissue: specialized plant tissue composed of tubelike cells that transports water and nutrients in some plants. (p. 300)

virus: a strand of DNA or RNA surrounded by a layer of protein that can infect and replicate in a host cell. (p. 247)

vacuna: mezcla que contiene material de uno o más patógenos desactivados, como los virus. (pág. 252)

tejido vascular: tejido especializado de la planta compuesto de células tubulares que transportan agua y nutrientes en algunas plantas. (pág. 300)

virus: filamento de ADN o de ARN rodeado por una capa de proteína que puede infectar una célula huésped y replicarse en ella. (pág. 247)

X

xylem (ZI lum): a type of vascular tissue that carries water and dissolved nutrients from the roots to the stem and the leaves. (p. 314)

xilema: tipo de tejido vascular que transporta agua y nutrientes disueltos desde las raíces hacia el tallo y las hojas. (pág. 314)

Z

zygosporangia (zi guh spor AN jee uh): tiny stalks formed when a zygote fungus undergoes sexual reproduction. (p. 279)

zigosporangia: tallos diminutos que se forman cuando un hongo zigoto se somete a reproducción sexual. (pág. 279)

Index

A

Academic Vocabulary, 270, 300, 336.
   See also **Vocabulary**
Active virus(es), 248
Aerobic bacteria, 233
Algae
   explanation of, *266,* **266**
   importance of, 269, 275
   types of, 268, *268*
Alternation of generations, *352,* **352,**
   353
Amoeba
   explanation of, *271,* **271**
   movement of, 271 *lab*
Angiosperm(s)
   explanation of, *319,* 319–320, *320,*
    356
   leaves of, 317
   life cycle of, 357, *357*
   types of, 319, *319*
Animal cell(s), *297, 297*
Animal-like protists
   explanation of, 266, *266*
   protozoans as, 270, *272, 272*
   types of, *270,* 270–271, *271*
Animals
   bacteria living in, 239, *239*
Annual plant(s), 320
Antibiotic(s)
   explanation of, **242,** 251
   from fungi, 281, 283, *283*
   resistance to, 242–243, *243,* 283
Antibody(ies)
   explanation of, **251**
   function of, 252 *lab*
Archaea, 231, 235
Ascus, 279
Asexual reproduction
   explanation of, 234, **266**
   in fungi, 278, 279
   in plants, 351, *351*
Athlete's foot, 280
Autotroph(s), 284
Auxin(s), 345, *345*

B

Bacteria
   aerobic, 233
   affecting investigations, 245
   antibiotic resistance by, 283
   archaea v., 235
   beneficial, *239,* 239–241, *240, 241*
   effect of disinfectants on, 254–255
    lab
   endospores and, 235, *235*
   in environment, 239 *lab*
   explanation of, **231**

food and, 241
   function of, 233
   harmful, *242,* 242–243, *243*
   method to kill, 237
   movement of, 234, *234*
   photosynthetic, 284
   reproduction of, 234, *234*
   resistant to antibiotics, 242–243
   size and shape of, 231 *lab,* 232, *232*
   slime layer in, 233 *lab*
   structure of, 232
Basidia, 278
Basidiocarp, 278, *278*
Biennial plant(s), 320
Big Idea, 228, 256, 262, 288, 294, 324,
   330, 362
   Review, 259, 291, 327, 365
Bioremediation, 241
Blue cheese, 280
Bryophyte(s), 307, *307*

C

Cambium, 314
Cancer
   research on, 252
Carbon dioxide
   deforestation and, 339
   plant use of, 333, 334, 337
Cattle
   bacteria living in, 239, *239*
Cell division, 352, 353
Cellular respiration
   explanation of, **336**
   observation of, 336 *lab*
   photosynthesis compared to, 337,
   337
Cellulose, 299
Chapter Review, 258–259, 290–291,
   326–327, 364–365
Chemicals
   response of plants to, *345,* 345–346,
   346
Chlorophyll
   in plants, 335
Chloroplast(s), *297,* 334, *334,* 335
Cilia, 270
Ciliate(s), 270
Climate change
   greenhouse gases and, 339
Clostridium botulinum, 242
Club fungi, 278, *278*
Club moss(es), 309, *309*
Common Use. *See* **Science Use v.**
   Common Use
Conifer(s) , 318, *318,* 355
Conjugation, *234,* **234,** 270
Cotyledons, 320

Critical thinking, 236, 244, 253, 259,
   274, 285, 291, 304, 310, 321, 338,
   348, 359
Cuticle, 299 *lab,* 317, 334, *334*
Cycad(s), 318
Cystic fibrosis
   research on, 252
Cytokinin(s), 346, *346*
Cytoplasm
   in bacteria, 232, *232*

D

Decomposer(s)
   fungi as, 281, *281*
Decomposition
   in aerobic and anaerobic environ-
    ments, 240 *lab*
   explanation of, **240**
Deforestation
   carbon dioxide and, 339, *339*
Diatom, *267,* **267**
Dicot(s), 320, *320*
Diffusion, 300, 307
Dinoflagellates, 267, *267*
Diploid cell(s), 352
Diploid generation, 352, 353, 355
Disease(s)
   bacterial, *242,* 242–243, *243*
   from fungi, 279, 283
   protists as cause of, 272
   viral, 247, *250,* 250–252, *251*
Disinfectant(s)
   bacterial growth and, 254–255 *lab*
DNA
   in bacteria, 232, 234, *234*
   in viruses, 247, 249, 252

E

Ecosystem(s)
   role of algae in, 269, *269*
   role of fungus-like protists in, 273
Endospore(s)
   explanation of, **235**
   formation of, *235*
Energy
   explanation of, 336
   in light, 335, *335,* 337, *337*
Environment
   response of plants to, *342,* 342–344,
   343, 344
Epidermal tissue
   in leaves, 317
Ethylene, 345, *345*
Euglena, 265
Euglenoid(s), 267, *267*
Eukaryote(s)
   fungi as, 277
   protists as, 265, 266

Eukaryotic cell(s), 297
Evergreen(s), 355
Extremophile(s), 235

F

Fern(s)
  explanation of, 309, *309*
  life cycle of, 353, *353*
Fertilization, 352
Fission, 234
Flagella, *234*, **234**, 267, 270
Flagellate(s), 270
Fleming, Alexander, 283
Flower(s)
  explanation of, 356, *356*
  modeling, 356 *lab*
  plants that product, 319, *319*, 320
  as response to darkness, 344, *344*,
    345 *lab*
Flowering seed plants(s)
  explanation of, 354
  reproduction in, 356, 356–358, *357*,
    358
Flowerless seed plant(s)
  explanation of, 354
  reproduction in, 355, *355*
Foldables, 231, 241, 248, 257, 272, 280,
  289, 298, 308, 318, 325, 334, 342,
  354, 363
Food
  bacteria and, 241
Food poisoning, 243
Frond(s), **309**
Fruit(s)
  explanation of, **357**
  as food source, 358
  identification of, 351 *lab*
Fungi
  classification of, 278
  club, 278, *278*
  examination of, 277 *lab*
  explanation of, 277, *277*, 288
  illness related to, 283
  imperfect, 280
  importance of, 281, *281*
  lichens and, 284, *284*
  medical uses for, 283, *283*
  plant roots and, 282, *282*
  sac, 279, *279*
  zygote, 279, *279*
Funguslike protist(s)
  explanation of, 266, *266*
  importance of, 273
  slime and water molds as, 273, *273*

G

Gene transfer
  research on, 252
Generations, **352**
Genetic disorder(s)
  viruses to treat, 252
Gibberellin(s), 346, *346*
Ginkgo, 318
Gravity
  response of plants to, 343, *343*

Great Irish Potato Famine, 273
Green Science, 275, 339
Greenhouse gas(es)
  climate change and, 339
Gymnosperm(s)
  explanation of, 318, *318*, 355, *355*
  leaves of, 317

H

Haploid cell(s), 352
Haploid generation, 352
Haploid spore(s), 353
Harmful algal bloom (HAB), 269
Hemophilia
  research on, 252
Heterotroph(s), 277
HIV (human immunodeficiency
    virus), 250, *250*
Hormone(s)
  plant, 345–346, *346*, *347*
Hornwort(s), 308, *308*
Horsetail(s), 309, *309*
How It Works, 237
How Nature Works, 305
Human(s)
  plant responses and, 346, *347*
Hyphae
  explanation of, **277**, *277*, 282, 284
  of zygote fungus, 279
Hypochlorite, 237

I

Immunity, **251**
Imperfect fungi, 280
Influenza, 250, *250*
Interpret Graphics, 236, 244, 253, 274,
  285, 304, 310, 321, 338, 348, 359

K

Kelp forest(s), 269, *269*
Kernel(s), 358
Key Concepts, 230, 238, 246, 264, 276,
  296, 306, 312, 332, 340, 350
  Check, 231, 233, 240, 242, 243, 248,
    252, 266, 269, 272, 273, 278, 283,
    284, 298, 300, 301, 309, 313, 319,
    320, 333, 335, 336, 337, 343, 352,
    353, 354
  Summary, 256, 288, 324, 362–363
  Understand, 236, 244, 253, 258, 274,
    285, 290, 304, 310, 321, 338, 348,
    359, 364

L

Lab, 254–255, 286–287, 322–323,
  360–361. *See also* **Launch Lab;**
  MiniLab; Skill Practice
Lactobacillus, 239
Launch Lab, 231, 239, 247, 265, 277,
  297, 307, 313, 333, 341, 351
Leaves
  explanation of, 316 *lab*, 317, *317*
  photosynthesis and, 334, *334*

Lesson Review, 236, 244, 253, 274, 285,
  304, 310, 321, 338, 348, 359
Lichen(s)
  examination of, 286–287 *lab*
  explanation of, **284**
  importance of, 284
Light
  energy in, 335, *335*, 337, *337*
  response of plants to, 341, *341*, *342*
  response of seeds to, 349
Lignin, **299**
Lipid(s), 275
Liverwort(s), 301, *301*, 308, *308*

M

Malaria
  spread of, 272, *272*
Mangrove forest(s), 305, *305*
Math Skills, 235, 236, 259, 282, 291,
  319, 321, 327, 346, 365
Medicine
  from fungi, 283, *283*
Meiosis, **352**
Mesophyll, 334, *334*
Microalgae, 275
MiniLab, 233, 240, 252, 271, 280, 299,
  316, 336, 345, 356. *See also* **Lab**
Mitosis, 352, **353**
Monocot(s), 320, *320*
Mosquito(es)
  malaria spread by, 272
Moss(es)
  explanation of, 301, *301*
  life cycle of, 353, *353*
  role of, 308, *308*
  water retention by, 307 *lab*, 308
Multicellular, **298**
Mushroom(s)
  explanation of, **278**
  as fungi, 277, *277*
  poisonous, 283
Mutation(s), **249**
Mycelium, **277**, *277*
Mycorrhiza, **282**, 284

N

Nitrogen fixation, *240*, **240**
Nonvascular seedless plant(s), 301,
  307, 307–308, *308*

O

Oil
  algae and, 275
Organelle(s), **297**, *297*
Osmosis, 300, **307**
Ovary(ies)
  in flowers, **356**
Ovule(s), **354**
Oxygen
  from photosynthesis, 275
  in plants, 333, 334, 335, *335*, 337

SCIENCE SKILL HANDBOOK

MATH SKILL HANDBOOK

REFERENCE HANDBOOK

GLOSSARY/ GLOSARIO

INDEX

P

Palisade mesophyll cell(s), 317, *317,* 334, *334*
Paramecium, 266, *266, 270,* **270**
Parasite(s)
  fungi as, 277
  protozoans as, 272
Pasteurization, 243
Pathogen(s)
  affecting investigations, 245
  explanation of, **242**
Peat moss, 307 *lab*
Penicillium, 283, *283*
Perennial plant(s), 320
Phloem, 333, *315,* **315**
Photobioreactor(s), 275
Photoperiodism, 344, 345 *lab*
Photosynthesis
  algae and, 265, *265,* 269, 275
  cellular respiration compared to, 337, *337*
  explanation of, 317, **334**
  food production through, 266
  function of, 282
  importance of, 335
  leaves and, 334, *334*
  lichens and, 284, *284*
  observation of, 336 *lab*
  process of, 335, *335,* 339
Phyla, 301
Pistil, 356
Plant(s)
  adaptations of, 298–300, *299, 300,* 322–323 *lab*
  characteristics of, *297,* 297–298, *298,* 311
  classification of, 301, *302–303*
  day-neutral, 344, *344*
  design of stimulating environment for, 360–361 *lab*
  environment and structure of, 311
  explanation of, 297 *lab*
  in extreme environments, 322–323 *lab*
  fungi and, 282, *282*
  long-day, 344, *344*
  movement of materials inside, 333, 333 *lab*
  in salt water, 305, *305*
  seed, 301, *313,* 313–320, 313 *lab,* 314, *315, 316, 317, 318, 319, 320*
  seedless, 301, *307,* 307–309, *308, 309*
  short-day, 344, *344*
  water in, 299 *lab,* 307 *lab*
Plant cell(s), 297, *297,* 299
Plant hormone(s)
  effects of, 346, *347*
  explanation of, **345**
Plant process(es)
  cellular respiration as, 336–337, *337*
  materials for, 333
  observation of, 336 *lab*
  photosynthesis as, *334,* 334–335, *335*
Plant reproduction
  alternation of generations and, 352, *352*
  in seedless plants, 353, *353*
  in seed plants, *354,* 354–358, *355, 356, 357, 358*
  types of, 351
Plant response(s)
  to chemical stimuli, *345,* 345–346, *346*
  effect of humans on, 346, *347*
  to environmental stimuli, *342,* 342–344, *343, 344*
  to stimuli, 341, *341,* 341 *lab*
Plantlike protist(s)
  algae as, 266, *266,* 269, *269*
  explanation of, 266, *266*
  types of, 267–268
Plasmodium(ia)
  explanation of, 272
  life cycle of, *272*
Pneumonia
  bacterium that causes, 232
Pod(s), 358
Pollen grain(s), *354,* **354**
Pollination, 354
Pollution
  bacteria that eat, 241
Process, 270
Producer(s), 298
Prokaryote(s)
  bacteria as, 231
  explanation of, 235
Protist(s)
  animal-like, *270,* 270–272, *271, 272*
  characteristics of, 265 *lab*
  classification of, 266, *266*
  explanation of, *265,* **265,** 288
  fungus-like, 273, *273*
  plant-like, 266–269, *267, 268, 269*
  reproduction of, 266
Protozoan(s)
  explanation of, **270**
  importance of, 272, *272*
Pseudopod(s), *271,* **271**

Q

Quercus rubra, 301

R

Reading Check, 234, 241, 243, 249, 250, 265, 267, 269, 270, 278, 279, 280, 282, 297, 299, 307, 314, 315, 320, 335, 341, 342, 344, 346, 351, 355, 357, 358
Red tide, 269, *269*
Replication
  of viruses, 247 *lab,* 248, *248–250,* 249
Reproduction. *See also* **Asexual reproduction; Sexual reproduction**
  asexual, 266
  in bacteria, 234
  in plants, 300, *300*
  in seedless plants, 353, *353*
  in seed plants, *354,* 354–358, *355, 356, 357, 358*
  sexual, 266
Resistant, 242, 243, *243*

Review Vocabulary, 249, 266, 307, 353. *See also* **Vocabulary**
Rhizoid(s), 307, 308
Ribosome(s)
  of archaea, 235
  explanation of, 232
RNA
  in viruses, 247, 249, 252
Root(s), 315, *315,* 316 *lab*

S

Sac fungi, 279
Salt water
  plants in, 305, *305*
Sarcodine(s), 271
Science Methods, 255, 287, 323, 361
Science Use v. Common Use, 242, 278, 314, 352. *See also* **Vocabulary**
Seed(s)
  dispersal of, 300, *300*
  explanation of, *354,* **354**
  as food source, 358
  in gymnosperms, 318, *318*
  response to light intensity, 349
Seed plant(s)
  angiosperms as, *319,* 319–320, *320*
  characteristics of, 313, 313 *lab*
  explanation of, 301
  gymnosperms as, 318, *318*
  leaves of, 317, *317*
  reproduction in, *354,* 354–358, *355, 356, 357, 358*
  roots of, 315, *315*
  stems of, 316, *316*
  types of, 354
  vascular tissue in, *314,* 314–315
Seedless plant(s)
  explanation of, 301
  nonvascular, *307,* 307–308, *308*
  reproduction in, 353, *353*
  vascular, 309, *309*
Sexual reproduction
  in fungi, 279
  offspring of, 266
  in plants, 351
  in seed plants, *354,* 354–358, *355, 356, 357, 358*
Silica
  in diatoms, 267, *267*
Skill Practice, 245, 311, 349. *See also* **Lab**
Slash-and-burn, 339
Slime mold(s), 266, *266,* 273, *273*
Sperm, 352
Spongy mesophyll cell(s), 334, *334,* 317, *317*
Spore(s)
  examination of, 280 *lab*
  explanation of, 278, **352**
  in sac fungi, 279
Stamen, 356
Standardized Test Practice, 260–261, 292–293, 328–329, 366–367
Stem(s), 316, *316,* 316 *lab*
Stimuli, 341
Stomata, 317, 334

Streptococcus pneumonia, 242
Study Guide, 256–257, 288–289, 324–325, 362–363
Sugar(s)
  in cellular respiration, 336
  in photosynthesis, 335, *335*

Tendril(s), 343, *343*
Thigmotropism, 343, *343*
Thrush, 283
Touch
  response of plants to, 343, *343*
Tracheid(s), *314,* **314**
Transport, 300
Tree(s)
  atmospheric carbon dioxide and, 339
Triglyceride(s)
  oil products from, 275
Tropism, 342
Tuberculosis, 242, *242*

Uranium
  bacteria that clean up contamination from, 241, *241*

Vaccine(s), 252
Vascular seedless plant(s)
  explanation of, 301, 309, *309*
Vascular tissue
  explanation of, **300**
  in seedless plants, 309
  in seed plants, *314,* 314–315, *315*
Vessel element(s), 314, *314*
Viral disease(s)
  explanation of, 250, *250*
  immunity to, 251, *251*
  medications to treat, 251
  types of, 247, 250
  vaccines to prevent, 252
Virus(es)
  beneficial, 252
  explanation of, **247,** 249
  latent, 248
  mutations in, 249
  organisms and, 248
  replication of, 247 *lab,* 248, *248–249,* 249
  research with, 252
  shapes of, 247, *247*
Visual Check, 232, 234, 239, 242, 248, 250, 251, 271, 278, 283, 301, 307, 317, 320, 334, 337, 343, 344, 354, 357, 358
Vocabulary, 229, 230, 238, 246, 256, 263, 264, 276, 288, 296, 306, 312, 332, 340, 350, 362. *See also*
  Academic Vocabulary; Review Vocabulary; Science Use v. Common Use; Word Origin
  Use, 236, 244, 253, 257, 274, 285, 289, 304, 310, 321, 325, 338, 348, 359, 363
Volvox, 268, *268*

Water mold(s), 273
Water
  in plants, 333, *333,* 337
What do you think?, 229, 236, 244, 253, 263, 274, 284, 295, 304, 310, 321, 331, 338, 348, 359
Word Origin, 231, 242, 251, 267, 284, 298, 314, 334, 342, 352. *See also*
  Vocabulary
Writing In Science, 259, 291, 327, 365

Xylem, 333, *314,* **314**

Yeast
  as sac fungi, 279

Zygosporangia, 279, *279*
Zygospore(s), 279
Zygote fungi, 279
Zygote, 352, 355

SCIENCE SKILL HANDBOOK

MATH SKILL HANDBOOK

REFERENCE HANDBOOK

GLOSSARY/ GLOSARIO

INDEX

Credits

Photo Credits

PERIODIC TABLE OF THE ELEMENTS

Element — Hydrogen
Atomic number — 1
Symbol — H
Atomic mass — 1.01
State of matter

- 🎈 Gas
- 💧 Liquid
- ▢ Solid
- ⊙ Synthetic

A column in the periodic table is called a **group.**

A row in the periodic table is called a **period.**

1								
1 Hydrogen 1 **H** 🎈 1.01	**2**							
Lithium 3 **Li** ▢ 6.94	Beryllium 4 **Be** ▢ 9.01	**3**	**4**	**5**	**6**	**7**	**8**	**9**
Sodium 11 **Na** ▢ 22.99	Magnesium 12 **Mg** ▢ 24.31							
Potassium 19 **K** ▢ 39.10	Calcium 20 **Ca** ▢ 40.08	Scandium 21 **Sc** ▢ 44.96	Titanium 22 **Ti** ▢ 47.87	Vanadium 23 **V** ▢ 50.94	Chromium 24 **Cr** ▢ 52.00	Manganese 25 **Mn** ▢ 54.94	Iron 26 **Fe** ▢ 55.85	Cobalt 27 **Co** ▢ 58.93
Rubidium 37 **Rb** ▢ 85.47	Strontium 38 **Sr** ▢ 87.62	Yttrium 39 **Y** ▢ 88.91	Zirconium 40 **Zr** ▢ 91.22	Niobium 41 **Nb** ▢ 92.91	Molybdenum 42 **Mo** ▢ 95.96	Technetium 43 **Tc** ⊙ (98)	Ruthenium 44 **Ru** ▢ 101.07	Rhodium 45 **Rh** ▢ 102.91
Cesium 55 **Cs** ▢ 132.91	Barium 56 **Ba** ▢ 137.33	Lanthanum 57 **La** ▢ 138.91	Hafnium 72 **Hf** ▢ 178.49	Tantalum 73 **Ta** ▢ 180.95	Tungsten 74 **W** ▢ 183.84	Rhenium 75 **Re** ▢ 186.21	Osmium 76 **Os** ▢ 190.23	Iridium 77 **Ir** ▢ 192.22
Francium 87 **Fr** ▢ (223)	Radium 88 **Ra** ▢ (226)	Actinium 89 **Ac** ▢ (227)	Rutherfordium 104 **Rf** ⊙ (267)	Dubnium 105 **Db** ⊙ (268)	Seaborgium 106 **Sg** ⊙ (271)	Bohrium 107 **Bh** ⊙ (272)	Hassium 108 **Hs** ⊙ (270)	Meitnerium 109 **Mt** ⊙ (276)

The number in parentheses is the mass number of the longest lived isotope for that element.

Lanthanide series	Cerium 58 **Ce** ▢ 140.12	Praseodymium 59 **Pr** ▢ 140.91	Neodymium 60 **Nd** ▢ 144.24	Promethium 61 **Pm** ⊙ (145)	Samarium 62 **Sm** ▢ 150.36	Europium 63 **Eu** ▢ 151.96
Actinide series	Thorium 90 **Th** ▢ 232.04	Protactinium 91 **Pa** ▢ 231.04	Uranium 92 **U** 🎈 238.03	Neptunium 93 **Np** ⊙ (237)	Plutonium 94 **Pu** ⊙ (244)	Americium 95 **Am** ⊙ (243)